CALCULUS
REORDERED

CALCULUS REORDERED

A History of the Big Ideas

DAVID M. BRESSOUD

PRINCETON UNIVERSITY PRESS
PRINCETON AND OXFORD

Published by Princeton University Press
41 William Street, Princeton, New Jersey 08540

In the United Kingdom: Princeton University Press
6 Oxford Street, Woodstock, Oxfordshire, OX20 1TR

Library of Congress Cataloging-in-Publication Data

ISBN 978-0-691-18131-8
LCCN 2018957493

British Library Cataloging-in-Publication Data is available

Editorial: Vickie Kearn and Lauren Bucca
Production Editorial: Sara Lerner
Text and Jacket Design: Carmina Alvarez
Production: Erin Suydam
Publicity: Sara Henning-Stout and Kathryn Stevens
Copyeditor: Jennifer Harris
Jacket Credit: A page from Sir Isaac Newton's
Waste Book, c. 1612-c. 1653.
From the Portsmouth Collection,
donated by the fifth Earl of Portsmouth, 1872.
Cambridge University Library.

This book has been composed in LaTeX

Printed on acid-free paper. ∞

press.princeton.edu

Printed in the United States of America

1   3   5   7   9   10   8   6   4   2

dedicated to Jim Smoak

for your inspirational love of mathematics and its history

Contents

Preface

This book will not show you how to do calculus. My intent is instead to explain how and why it arose. Too often, its narrative structure is lost, disappearing behind rules and procedures. My hope is that readers of this book will find inspiration in its story. I assume some knowledge of the tools of calculus, though, in truth, most of what I have written requires little more than mathematical curiosity.

Most of those who have studied calculus know that Newton and Leibniz "stood on the shoulders of giants" and that the curriculum we use today is not what they handed down over 300 years ago. Nevertheless, it is disturbingly common to hear this subject explained as if it emerged fully formed in the late seventeenth century and has changed little since. The fact is that the curriculum as we know it today was shaped over the course of the nineteenth century, structured to meet the needs of research mathematicians. The progression we commonly use today and that AP Calculus has identified as the *Four Big Ideas* of calculus—*limits, derivatives, integrals,* and finally *series*—is appropriate for a course of analysis that seeks to understand all that can go wrong in attempting to use calculus, but it presents a difficult route into *understanding* calculus. The intent of this book is to use the historical development of these four big ideas to suggest more natural and intuitive routes into calculus.

The historical progression began with integration, or, more properly, accumulation. This can be traced back at least as far as the fourth century BCE, to the earliest explanation of why the area of a circle is equal to that of a triangle whose base is the circumference of the circle ($\pi \times$ diameter) and whose height is the radius.[1] In the ensuing centuries, Hellenistic philosophers became adept at deriving formulas for areas and volumes by imagining geometric objects as built from thin slices. As we will see, this approach was developed further by Islamic, Indian, and Chinese philosophers, reaching its apex in seventeenth-century Europe.

Accumulation is more than areas and volumes. In fourteenth century Europe, philosophers studied variable velocity as the rate at which distance is changing at each instant. Here we find the first explicit use of accumulating small changes in distance to find the total distance that is traveled. These philosophers realized that if the velocity is represented by distance above a horizontal axis, then the area between the curve representing velocity and the horizontal axis corresponds to the distance that has been traveled. Thus, accumulation of distances can be represented as an accumulation of area, connecting geometry to motion.

The next big idea to emerge was differentiation, a collection of problem-solving techniques whose core idea is ratios of change. Linear functions are special because the ratio of the change in the output to the change in the input is constant. In the middle of the first millennium of the Common Era, Indian astronomers discovered what today we think of as the derivatives of the sine and cosine as they explored how changes in arc length affected changes in the corresponding lengths of chords. They were exploring *sensitivity*, one of the key applications of the derivative: understanding how small changes in one variable will affect another variable to which it is linked.

In seventeenth-century Europe, the study of ratios of change appeared in the guise of tangent lines. Eventually, these were connected to the general study of rates of change. Calculus was born when Newton, and then independently Leibniz, came to realize that the techniques for solving problems of accumulation and ratios of change were inverse to each other, thus enabling natural philosophers to use solutions found in one realm to answer questions in the other.

The third big idea to emerge was that of series. Though written as infinite summations, infinite series are really limits of sequences of partial sums. They arose independently in India around the thirteenth century and Europe in the seventeenth, building from a foundation of the search for polynomial approximations. By the time calculus was well established, in the early eighteenth century, series had become indispensable tools for the modeling of dynamical systems, so central that Euler, the scientist who shaped eighteenth-century mathematics and established the power of calculus, asserted that any study of calculus must begin with the study of infinite summations.

The term *infinite summation* is an oxymoron. "Infinite" literally means without end. "Summation," related to "summit," implies bringing to a conclusion. An infinite summation is an unending process that is brought to a conclusion. Applied without care, it can lead to false conclusions and apparent contradictions. It was largely the difficulties of understanding these infinite summations that led, in the nineteenth century, to the development of the last of our big ideas, the limit. The common use of the word "limit" is loaded with connotations that easily lead students astray. As Grabiner has documented,[2] the modern meaning of limits arose from the algebra of inequalities, inequalities that bound the variation in the output variable by controlling the input.

The four big ideas of calculus in their historical order, and therefore our chapter headings, are

(1) Accumulation (Integration)
(2) Ratios of Change (Differentiation)
(3) Sequences of Partial Sums (Series)
(4) Algebra of Inequalities (Limits).

In addition, I have added a chapter on some aspects of nineteenth-century analysis. Just as no one should teach algebra who is ignorant of how it is used in calculus, so no one should teach calculus who has no idea how it evolved in the nineteenth century. While strict adherence to this historical order may not be necessary, anyone who teaches calculus must be conscious of the dangers inherent in departing from it.

How did we wind up with a sequence that is close to the reverse of the historical order: limits first, then differentiation, integration, and finally series? The answer lies in the needs of the research mathematicians of the nineteenth century who uncovered apparent contradictions within calculus. As set out by Euclid and now accepted as the mathematical norm, a logically rigorous explanation begins with precise definitions and statements of the assumptions (known in the mathematical lexicon as *axioms*). From there, one builds the argument, starting with immediate consequences of the definitions and axioms, then incorporating these as the building blocks of ever more complex propositions and theorems. The beauty of this approach is that it facilitates the checking of any mathematical argument.

This is the structure that dictated the current calculus syllabus. The justifications that were developed for both differentiation and integration rested on concepts of limits, so logically they should come first. In some sense, it now does not matter whether differentiation or integration comes next, but the limit definition of differentiation is simpler than that of accumulation, whose precise explication as set by Bernhard Riemann in 1854 entails a complicated use of limit. For this reason, differentiation almost always follows immediately after limits. The series encountered in first-year calculus are, for all practical purposes, Taylor series, extensions of polynomial approximations that are defined in terms of derivatives. As used in first-year calculus, they could come before integration, but the relative importance of these ideas usually pushes integration before series.

The progression we now use is appropriate for the student who wants to verify that calculus is logically sound. However, that describes very few students in first-year calculus. By emphasizing the historical progression of calculus, students have a context for understanding how these big ideas developed.

Things would not be so bad if the current syllabus were pedagogically sound. Unfortunately, it is not. Beginning with limits, the most sophisticated and difficult of the four big ideas, means that most students never appreciate their true meaning. Limits are either reduced to an intuitive notion with some validity but one that can lead to many incorrect assumptions, or their study devolves into a collection of techniques that must be memorized.

The next pedagogical problem is that integration, now following differentiation, is quickly reduced to antidifferentiation. Riemann's definition of the integral—a product of the late nineteenth century that arose in response to the question of how discontinuous a function could be yet still be integrable—is difficult to comprehend, leading students to ignore the integral as a limit and focus on the integral as antiderivative. Accumulation is an intuitively simple idea. There is a reason this was the first piece of calculus to be developed. But students who think of integration as primarily reversing differentiation often have trouble making the connection to problems of accumulation.

The current curriculum is so ingrained that I hold little hope that this book will cause everyone to reorder their syllabi. My desire is that teachers and students will draw on the historical record to focus on the algebra of inequalities when studying limits, ratios of change when studying differentiation, accumulation when studying integration, and sequences of partial sums when studying series. To aid in this, I have included an appendix of practical insights and suggestions from research in mathematics education. I hope that this book will help teachers recognize the conceptual difficulties inherent in the definitions and theorems that were formulated in the nineteenth century and incorporated into the curriculum during the twentieth. These include the precise definitions of limits, continuity, and convergence. Great mathematicians did great work without them. This is not to say that they are unimportant. But they entered the world of calculus late because they illuminate subtle points that the mathematical community was slow to understand. We should not be surprised if beginning students also fail to grasp their importance.

I also want to say a word about how I refer to the people involved in the creation of calculus. Before 1700, I refer to them as "philosophers" because that is how they thought of themselves, as "lovers of wisdom" in all its forms. None restricted themselves purely to the study of mathematics. Newton and Leibniz are in this company. Newton referred to physics as "natural philosophy," the study of nature. From 1700 to 1850, I refer to them as "scientists." Although that word would not be invented until 1834, it accurately captures the broad interests of all those who worked to develop calculus during this period. Many still considered themselves to be philosophers, but the emphasis had shifted to a more practical exploration of the world around us. Almost all of them included an interest in astronomy and what today we would call "physics." After 1850, it became common to focus exclusively on questions of mathematics. In this period and only in this period, I will refer to them as mathematicians.

I owe a great debt to the many people who have helped with this book. Jim Smoak, a mathematician without formal training but a great knowledge of its history, helped to inspire it, and he provided useful feedback on a very early draft. I am indebted to Bill Dunham and Mike Oehrtman who gave me many helpful suggestions. Both Vickie Kearn at Princeton University Press and Katie Leach at Cambridge University Press expressed an early interest in this project. Their encouragement helped spur me to complete it. Both sent my first draft out to reviewers. The feedback I received has been invaluable. I especially wish to thank the Cambridge reviewer who went through that first draft line by line, tightening my prose and suggesting many cuts and additions. You will see your handiwork throughout this final manuscript. I want to thank my production editor, Sara Lerner, and especially my copyeditor, Glenda Krupa. Finally, I want to thank my wife, Jan, for her support. Her love of history has helped to shape this book.

David M. Bressoud
bressoud@macalester.edu
August 7, 2018

CALCULUS
REORDERED

Chapter 1

ACCUMULATION

This chapter will follow the development of the most intuitive of the big ideas of calculus, that of accumulation. We begin with the discovery of formulas for areas and volumes by the Greek philosophers Antiphon, Democritus, Euclid, Archimedes, and Pappus. This leads to the development of formulas for volumes of revolution by al-Khwarizmi, Kepler, and a host of seventeenth-century philosophers. We then move back to the fourteenth century to the application of accumulation for finding distance when the velocity is known, sketching the contributions of the Mertonian scholars and Nicole Oresme. Back in the seventeenth century, we will share in the amazement that came with the discovery of objects of infinite length yet finite volume, we will see how to turn arc lengths into areas, and we will conclude with the uses that Galileo and Newton made of accumulation to solve the greatest scientific mystery of the age: how it is possible for the earth to travel through space at incredible speeds without our experiencing the least sense of its motion.

1.1
Archimedes and the Volume of the Sphere

In 1906, Johan Ludwig Heiberg discovered a previously unknown work of Archimedes, *The Method of Mechanical Theorems*, within a thirteenth-century prayer book. The Archimedean text, which had been copied from an earlier manuscript sometime in the tenth century, had been scraped off the vellum pages so that they could be reused. Fortunately, much of the original text was still decipherable. What was readable was published in the following decade. In 1998, an anonymous collector purchased the text for two million dollars and handed it over to the Walters Art Museum in Baltimore, which has since supervised its preservation and restoration as well as its decipherment using modern scientific tools.

Archimedes wrote the *Method*, as this book has come to be known, for his contemporary and colleague Eratosthenes. In it, he explained his methods for computing areas, volumes, and moments. This text lays out the core ideas of integral calculus, including the use of infinitesimals, a technique that Archimedes hid when he wrote his formal proofs. A 2003 *NOVA* program about this manuscript claimed that

> this is a book that could have changed the history of the world. . . . If his secrets had not been hidden for so long, the world today could be a very different place. . . . We could have been on Mars today. We could have accomplished all of the things that people are predicting for a century from now. (*NOVA*, 2003)

The implication is that if the world had not lost Archimedes' *Method* for those centuries, calculus would have been developed long before. That is nonsense. As we shall see, Archimedes' other works were perfectly sufficient to lead the way toward the development of calculus. The delay was not caused by an incomplete understanding of Archimedes' methods but by the need to develop other mathematical tools. In particular, scholars needed the modern symbolic language of algebra and its application to curves before they could make substantial progress toward calculus as we know it. The development of this language and its application to analytic geometry would not be accomplished until the early seventeenth century. Even then, it took several decades to transform the "method of exhaustion" into algebraic techniques for computing areas and volumes. The work of Eudoxus, Euclid, and Archimedes was essential in the development of calculus, but not all of it was necessary, and it was far from sufficient.

Archimedes of Syracuse (circa 287–212 BCE) was the great master of areas and volumes. Although we cannot be certain of the year of his birth, the year of his death is all too sure. Sicily had allied with Carthage during the Second Punic War (218–201 BCE), the war that saw Hannibal cross the Alps with his elephants to attack Rome. The Roman general Marcellus laid a two-year siege on Syracuse, then the capital of Sicily. Archimedes was a master engineer who helped defend the city with weapons he

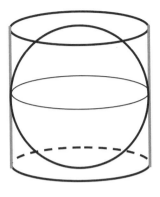

Figure 1.1. Sphere with the smallest cylinder that contains it.

invented: grappling hooks, catapults, and perhaps even mirrors to concentrate the sun's rays to burn Roman ships. Archimedes died during the sacking of the city when the Romans finally broke through the defenses. There is a story, possibly apocryphal, that General Marcellus tried to bring him to safety, but Archimedes was too engrossed in his mathematical calculations to follow.

Of his many accomplishments, Archimedes considered his greatest to be the formula for spherical volume—namely that the volume of a sphere is equal to two-thirds of the volume of the smallest cylinder that contains the sphere (see Figure 1.1). Archimedes valued this discovery so highly that he had a sphere embedded in a cylinder and the ratio 2:3 carved as his funeral monument, an object that still existed over a hundred years later when Cicero visited Syracuse.[1] To see why this gives us the usual formula for the volume of a sphere, let r be its radius. The smallest cylinder containing this sphere has a circular base of radius r and height $2r$, so its volume is

$$\text{volume of cylinder} = \pi(\text{Radius})^2(\text{Height}) = \pi r^2 \cdot 2r = 2\pi r^3.$$

Two-thirds of this is $(4/3)\pi r^3$, the volume of a sphere.

As Archimedes explained to Eratosthenes (with some elaboration on my part), he thought of the sphere as formed by rotating a circle around its diameter and imagined its volume as composed of thin slices perpendicular to the diameter. He began with a circle of diameter AB (Figure 1.2). Let X denote a point on this diameter and consider the perpendicular from X to the point C on the circle. If we rotate the area within the circle around the diameter AB, the thin slice perpendicular to the diameter at X is a disc of

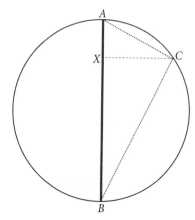

Figure 1.2. Circle with diameter AB.

area $\pi \overline{XC}^2$ and infinitesimal thickness ΔX. We represent the sum of the volumes of all of these discs as

$$\text{Volume of Sphere} = \sum \pi \overline{XC}^2 \, \Delta X.$$

Now Archimedes relied on some simple geometry. By the Pythagorean theorem, $\overline{XC}^2 = \overline{AC}^2 - \overline{AX}^2$. Because the angle $\angle ACB$ is a right angle, triangles AXC and ACB are similar. We obtain

$$\frac{\overline{AX}}{\overline{AC}} = \frac{\overline{AC}}{\overline{AB}}, \quad \text{or} \quad \overline{AC}^2 = \overline{AX} \cdot \overline{AB}.$$

Putting these together yields

$$\text{Volume of Sphere} = \sum \pi \overline{XC}^2 \, \Delta X$$
$$= \sum \pi \overline{AC}^2 \, \Delta X - \sum \pi \overline{AX}^2 \, \Delta X$$
$$= \sum \pi \overline{AX} \cdot \overline{AB} \, \Delta X - \sum \pi \overline{AX}^2 \, \Delta X.$$

The second summation is the volume of a cone. If we take our same diameter AB and at point X go out to a point D for which $\overline{AX} = \overline{AD}$, we get an isosceles right triangle (Figure 1.3). When we rotate that triangle around the axis AB, we get a cone of height $\overline{AB}$ with a base of radius

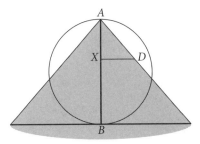

Figure 1.3. Circle with isosceles right triangle.

$\overline{AB}$. Its volume is equal to $\frac{1}{3}\pi\overline{AB}^3$ or, as Archimedes would have understood it, as $\frac{4}{3}$rds of the volume of the smallest cylinder that contains the sphere, the cylinder of height $\overline{AB}$ and radius $\frac{1}{2}\overline{AB}$. He had now established that

$$\text{Volume of Sphere} + \frac{4}{3}\text{Volume of Cylinder} = \sum \pi\overline{AX}\cdot\overline{AB}\,\Delta X.$$

The summation on the right-hand side is problematic as it stands. Archimedes neatly finished his derivation by considering moments. One use of moments is to determine balance. The moment is the product of mass and the distance from the pivot. Two objects of different masses on a seesaw can be in balance if their moments are equal, or, equivalently, if the ratio of their masses is the reciprocal of the ratio of their distances from the pivot (Figure 1.4). Archimedes was working with volumes, not masses, but if the densities are the same, then the ratio of the volumes equals the ratio of the masses. We take our two volumes on the left side of the equality and multiply them by $\overline{AB}$, effectively placing them at distance $\overline{AB}$ to the left of our pivot (Figure 1.5).

Multiplying the right side of our equality by $\overline{AB}$ yields

$$\sum \pi\overline{AX}\cdot\overline{AB}^2\,\Delta X.$$

Now $\pi\overline{AB}^2\,\Delta X$ is the volume of a disc of radius $\overline{AB}$ and thickness ΔX. Multiplying it by $\overline{AX}$ corresponds to the moment of such a disc at distance $\overline{AX}$ from the pivot. Adding up the moments of these discs gives us the moment of a fat cylinder of radius $\overline{AB}$ that rests along the balance beam from the pivot out to distance $\overline{AB}$ (Figure 1.5). Because this is a cylinder of constant

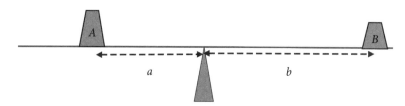

Figure 1.4. Weight A at distance a will balance weight B at distance b if $Aa = Bb$ or, equivalently, if $A/B = b/a$.

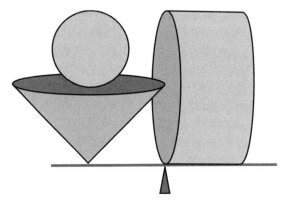

Figure 1.5. The sphere and the cone balance the fat cylinder.

radius, the total moment of all of these discs is the same as the moment were the fat cylinder to be placed at distance $\frac{1}{2}\overline{AB}$ from the pivot. The radius of the fat cylinder is $\overline{AB}$, twice the radius of the smallest cylinder that contains the sphere, so the volume of the fat cylinder is four times the volume of the cylinder that contains the sphere.

Now we can use the fact that the ratio of the volumes equals the ratio of the masses equals the reciprocal of the ratio of the distances from the pivot,

$$\frac{\text{Volume of Sphere} + \frac{4}{3}\text{Volume of Cylinder}}{4 \times \text{Volume of Cylinder}} = \frac{1}{2},$$

which gives us the result we seek,

$$\text{Volume of Sphere} = \frac{2}{3}\text{Volume of Cylinder}.$$

This argument was good enough to convince a colleague. It did not constitute a publishable proof. Archimedes would go on to supply such a proof in *On the Sphere and Cylinder*, but rather than trying to explain the intricacies of this technically challenging proof, I will illustrate the essence of the issues Archimedes faced in a much simpler example, that of demonstrating the formula for the area of a circle.

1.2
The Area of the Circle and the Archimedean Principle

Archimedes built on a technique that was much older. He credited the idea of using infinitely thin slices to find areas and volumes to Eudoxus of Cnidus who lived in the fourth century BCE on the southwest coast of what is today Turkey. Eudoxus had used this method of slicing to discover that the volume of a pyramid or cone is one-third the area of the base times the height. Even before Eudoxus, Antiphon of Athens (fifth century BCE) is credited with discovering that the area of a circle is equal to the area of a triangle with height equal to the radius of the circle and base given by the circumference of the circle.

In modern notation, we define π as the ratio of the circumference of a circle to its diameter,[2] so the circumference is π times the diameter, or $2\pi r$. The area of the triangle is half the height times the base, which is

$$\frac{1}{2}r \cdot 2\pi r = \pi r^2,$$

the familiar formula for the area of a circle. The formula emerges if we consider building a circle out of very thin triangles (see Figure 1.6). The triangles have heights that are close to the radius of the circle, and these heights approach the radius as the triangles get thinner. The sum of the bases of the triangles is close to the circumference of the circle, and again gets closer as the triangles get thinner. The total area of all of the triangles is the sum of half the base times the height, which is equal to half the sum of the bases times the height. This approaches half the circumference (the sum of the bases) times the radius.

Figure 1.6. A circle approximated by thin triangles.

What I now give is a slight paraphrasing and elaboration of Archimedes proof of the formula for the area of a circle. It relies on Proposition 1 from Book X of Euclid's *Elements*.

> Two unequal magnitudes being set out, if from the greater there is subtracted a magnitude greater than its half, and from that which is left a magnitude greater than its half, and if this process is repeated continually, then there will be left some magnitude less than the lesser magnitude set out. (Euclid, 1956, vol. 3, p. 14)

What this tells us is that if we have two positive quantities, leave one fixed and keep removing half from the other, then eventually (in a finite number of steps) the amount that remains of the quantity that has been successively halved will be less than the amount left unchanged. Today this is known as the *Archimedean Principle*, even though it goes back at least to Euclid. It may seem so obvious as not to be worth mentioning, but it should be noted that it explicitly rules out the possibility of an infinitesimal, a quantity that is larger than zero but smaller than any positive real number. If we allowed the fixed quantity to be an infinitesimal and the other to be a positive real number, then no matter how many times we take half of the real number, it will always be larger than the infinitesimal.

Theorem 1.1 (Archimedes, from *Measurement of a Circle*).   *The area of a circle is equal to the area of a right triangle whose height is the radius of the circle and whose length is the circumference.*

Proof. Following Archimedes' proof, we will demonstrate that the area of the circle is exactly equal to the area of the triangle by showing that it is neither smaller than the area of the triangle nor larger than the area of the

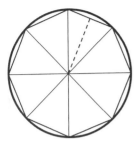

Figure 1.7. A circle with an inscribed octagon. The dashed line shows the height of one of the triangles.

Figure 1.8. Comparing the area between the circle and the first polygon to the area between the circle and the polygon with twice as many sides.

triangle. We first assume that A, the area of the circle, is strictly larger than T, the area of the triangle, i.e., that $A - T > 0$.

We consider an inscribed polygon, such as the octagon shown in Figure 1.7. We let P denote the area of the polygon. Because this polygon is inscribed in the circle, its area is less than that of the circle, $A - P > 0$. The area of the polygon is the sum of the areas of the triangles. Because each triangle has height less than the radius of the circle and the sum of the lengths of the bases of the triangles is less than the circumference of the circle, the area of the polygon is also less than the area of the triangle, $P < T$.

We now form a new polygon with twice as many sides by inserting a vertex on the circle exactly halfway between each pair of existing vertices. We label its area P'. I claim that $A - P'$ is less than half of $A - P$. To see why this is so, consider Figure 1.8. It is visually evident that the area that is filled by adding extra sides accounts for more than half of the area between the circle and the original polygon. We continue to double the number of sides until we get an inscribed polygon of area P^* for which $A - P^* < A - T$. The Archimedean principle promises us that this will happen eventually. When it does, then $P^* > T$.

But the polygon of area P^* is still an inscribed polygon, so $P^* < T$. Our assumption that the area of the circle is larger than T cannot be correct.

What if the area of the circle is strictly less than T? In that case, $T - A > 0$, and we let P be the area of a circumscribed polygon (see Figure 1.9). The height of each triangle that makes up our polygon is now equal to the radius,

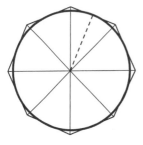

Figure 1.9. A circle with a circumscribed octagon. The dashed line shows the height of one of the triangles.

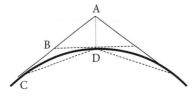

Figure 1.10. Comparing the area between the circle and a circumscribed polygon to the area between the circle and a circum-scribed polygon with twice as many sides.

but the perimeter of the polygon is strictly greater than the circumference of the circle, so $P > T$.

Once again we double the number of sides of the polygon by inserting a new vertex exactly halfway between each existing pair of vertices, and we let P' denote the area of the new polygon. Figure 1.10 shows how much of the area $P - A$ is removed when we double the number of sides. Because $BC = BD$, it follows that AB is more than half of AC. Comparing triangle ACD and BCD, they both have the same height (perpendicular distance from D to the line through AC) and the base of ACD is more than twice the base of BCD, it follows that doubling the number of sides takes away more than half of the area between the polygon and the circle, $P' - A < \frac{1}{2}(P - A)$.

We repeat this until $P^* - A < T - A$. This implies that $P^* < T$, contradicting the fact that every circumscribed polygon has an area greater than T. Because A can be neither strictly greater than T nor less than T, it must be exactly equal to T. □

The proof we have just seen may seem cumbersome and pedantic. Most people would be convinced by Figure 1.6. The problem is that such an argument relies on accepting "infinitely many" and "infinitely small" as meaningful quantities. Hellenistic philosophers *were* willing to use these as useful fictions that could help them discover mathematical formulas.

They were not willing to embrace them as sufficient to establish the validity of a mathematical result.

In the seventeenth century, philosophers engaged in heated debates over whether it was legitimate to derive results from nothing more than an analysis of infinitely thin slices. One sees in the work of both Newton and Leibniz a recognition of the power of arguments that rest on the use of infinitesimals, combined with a reluctance to abandon the rigor that Archimedes insisted upon. This reluctance would dissipate under the influence of the Bernoullis and Euler in the eighteenth century, but the problems this engendered would come roaring back in the early nineteenth in the form of apparent contradictions and paradoxes. In chapter 4, we will see how Cauchy recast the arguments of Archimedes and his Hellenistic successors into the precise language of limits in order to establish the modern foundations of calculus.

1.3
Islamic Contributions

In the centuries following Archimedes, mathematics declined as the Roman Empire grew. There never were many people who could read and understand the works of Euclid or Archimedes, much less build upon them. The continuation of their work required an unbroken chain of teachers and students steeped in these methods. For several centuries, Alexandria remained the one bright center of learning in the Eastern Mediterranean, but even there the number of teachers gradually declined.

One of the final flashes of mathematical brilliance occurred in the early fourth century CE with Pappus of Alexandria (circa 290–350 CE), the last great geometer of the Hellenistic world. His *Synagoge* or *Collection* was written as a commentary on and companion to the great Greek geometric texts that still existed in his time. In many cases, the original texts have since disappeared. Our knowledge of what they contained, even the fact of their existence, rests solely on what Pappus wrote about them. One of these lost books is *Plane Loci* by Apollonius of Perga (circa 262–190 BCE). Pappus preserved the statements of Apollonius's theorems, but not the proofs. As we shall see, these tantalizing hints of Hellenistic accomplishments would provide direct inspiration for Fermat, Descartes, and their contemporaries in the seventeenth century.

In the Greco-Roman world, virtually all mathematical work ceased in the late fifth century when the Musaeum of Alexandria—the Temple of the Muses—and its associated library and schools were suppressed because of their pagan associations.[3] All was not lost, however. The rise of the Abbasid empire in the eighth century would see renewed interest and significant new developments in mathematics.

Harun al-Rashid (763 or 766–809 CE) was the fifth Abbasid caliph or ruler. Stories of his exploits figure prominently in the classic tales of the *One Thousand and One Nights*. The Abbasids were descendants of the Prophet Muhammad's youngest uncle, and they took control of most of the Islamic world in 750. In 762 they moved their capital from Damascus to Baghdad. Among al-Rashid's supreme accomplishments was the founding of the Bayt al-Hikma or House of Wisdom. It was a center for the study of mathematics, astronomy, medicine, and chemistry. Its library collected and translated important scientific texts gathered from the Hellenistic Mediterranean, Persia, and India, and it ushered in a great flowering of Islamic[4] science that would last until the Mongol invasions of the thirteenth century.

Thabit ibn Qurra (836–901) was one of the scholars of the House of Wisdom who built on the work of both Greek and Islamic scholars. One of his accomplishments was the rediscovery of the formula for the volume of a paraboloid, the solid formed when a parabola is rotated about its main axis. Although this result had been known to Archimedes, there is every indication that ibn Qurra discovered it anew.

Cast into modern language, the derivation of this formula begins with recognition that a parabola is characterized as a curve for which the distance from the major axis is proportional to the square root of the distance along the major axis from the vertex. In modern algebraic notation, if the vertex is located at $(0,0)$ and x is the distance from the vertex, then y, the distance from the axis, can be represented by $y = a\sqrt{x}$ (Figure 1.11). The cross-sectional area of the paraboloid at distance x is $\pi \left(a\sqrt{x}\right)^2 = \pi a^2 x$. To approximate the volume over $0 \leq x \leq b$, we slice the paraboloid into n discs of thickness b/n. At $x = ib/n$, for each $0 \leq i < n$, the volume of the disc is

$$\pi a^2 \frac{ib}{n} \times \frac{b}{n} = \frac{\pi a^2 b^2}{n^2} i.$$

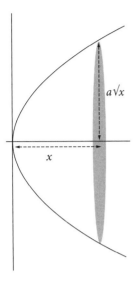

$a\sqrt{x}$

x

Figure 1.11. Cross-section of a paraboloid.

We now add the volumes of the individual discs,[5]

$$\frac{\pi a^2 b^2}{n^2}(0+1+2+\cdots+(n-1)) = \frac{\pi a^2 b^2}{n^2} \times \frac{n^2-n}{2} = \frac{\pi a^2 b^2}{2} - \frac{\pi a^2 b^2}{2n}.$$

As we take larger values of n (and thinner discs), the second term can be made as small as we wish, guaranteeing that the actual value can be neither smaller nor larger than $\pi a^2 b^2/2$.

Ibn al-Haytham (965–1039) demonstrated the power of this approach when he showed how to calculate the volume of the solid obtained by rotating this area about a line perpendicular to the axis of the parabola (Figure 1.12). If the parabolic curve is represented by $y = b\sqrt{x/a}$, where $0 \le y \le b$, then the radius of the disc at height ib/n is given by

$$a - \frac{ay^2}{b^2} = a - \frac{a(ib/n)^2}{b^2},$$

and the volume of the disc at height $y = ib/n$ is

(1.1) $\qquad \pi \left(a - \frac{a(ib/n)^2}{b^2}\right)^2 \times \frac{b}{n} = \pi a^2 b \left(\frac{1}{n} - \frac{2i^2}{n^3} + \frac{i^4}{n^5}\right).$

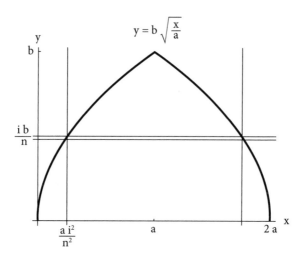

Figure 1.12. A vertical cross-section of al-Haytham's solid of revolution showing the horizontal slice.

It only remains to sum this expression over i from 1 to $n-1$. We need closed formulas for $1^2 + 2^2 + 3^2 + \cdots + (n-1)^2$ and $1^4 + 2^4 + 3^4 + \cdots + (n-1)^4$.

In his text *On Spirals*, Archimedes derived the formula for the sum of squares by showing that if

$$S(n) = (n+1)n^2 + (1+2+\cdots+n) = (n+1)n^2 + \frac{n(n+1)}{2},$$

then

$$S(n+1) - S(n) = 3(n+1)^2.$$

Since $S(1) = 3$, it follows that

$$S(n) = 3\left(1^2 + 2^2 + \cdots + n^2\right),$$

or, equivalently,

$$1^2 + 2^2 + \cdots + n^2 = \frac{(n+1)n^2}{3} + \frac{n(n+1)}{6}.$$

Abu Bakr al-Karaji (953–c. 1029) had discovered the formula for the sum of cubes,

$$1^3 + 2^3 + \cdots + n^3 = (1 + 2 + \cdots + n)^2 = \frac{n^2(n+1)^2}{4}.$$

Once he had guessed the formula, it was easy to verify by observing that the right side is 1 when $n = 1$, and the right side increases by $(n + 1)^3$ when n is replaced by $n + 1$.

Beyond the cubes, the problem gets harder because the formulas are not easy to guess. The genius of al-Haytham was to show how to use a known formula for the sum of the first n kth powers to find the formula for the sum of the first n $k + 1$st powers. He did this using specific sums, but his approach translates easily into a general statement. Seeking a formula for the sum of the first n $k + 1$st powers, we begin with

$$(n + 1)\left(1^k + 2^k + \cdots + n^k\right).$$

We distribute $n + 1$ through the sum, breaking it into two pieces so that $(n + 1)i^k$ becomes

$$(i + (n + 1 - i))\, i^k = i^{k+1} + (n + 1 - i)i^k.$$

It follows that

$$(1.2) \quad (n + 1)\left(1^k + 2^k + \cdots + n^k\right) = \left(1^{k+1} + 2^{k+1} + \cdots + n^{k+1}\right)$$

$$+ n \cdot 1^k + (n - 1)2^k + \cdots + 1 \cdot n^k$$

$$= \left(1^{k+1} + 2^{k+1} + \cdots + n^{k+1}\right)$$

$$+ \left(1^k + 2^k + \cdots + n^k\right)$$

$$+ \left(1^k + 2^k + \cdots + (n - 1)^k\right) +$$

$$+ \cdots + \left(1^k + 2^k\right) + 1^k.$$

The key to simplifying this relationship is the fact that the formula for the sum of the first n kth powers is of the form $n^{k+1}/(k + 1) + p_k(n)$ where p_k is a polynomial of degree at most k. As al-Haytham knew, this is true for $k = 1, 2$, and 3. The remainder of this derivation establishes that if it is true for the exponent k, then it holds for the exponent $k+1$. We make this substitution on both sides of equation (1.2).

$$(n+1)\left(\frac{n^{k+1}}{k+1}+p_k(n)\right)=\left(1^{k+1}+2^{k+1}+\cdots+n^{k+1}\right)+\frac{1}{k+1}$$

$$\left(n^{k+1}+(n-1)^{k+1}+\cdots+1^{k+1}\right)+p_k(n)$$

$$+p_k(n-1)+p_k(n-2)+\cdots+p_k(1)$$

$$\frac{n^{k+2}}{k+1}+\frac{n^{k+1}}{k+1}+np_k(n)+p_k(n)=\frac{k+2}{k+1}\left(1^{k+1}+2^{k+1}+\cdots+n^{k+1}\right)+p_k(n)$$

$$+p_k(n-1)+p_k(n-2)+\cdots+p_k(1).$$

Multiplying through by $(k+1)/(k+2)$ and solving for the sum of the $k+1$st powers, we get the desired relationship

(1.3)
$$1^{k+1}+2^{k+1}+\cdots+n^{k+1}=\frac{n^{k+2}}{k+2}+p_{k+1}(n),$$

where $p_{k+1}(n)$ is a polynomial in n of degree at most $k+1$.[6]

Now returning to the expression for the volume of each disc, equation (1.1), we can add these volumes:

$$\text{total volume}=\sum_{i=1}^{n}\pi a^2 b\left(\frac{1}{n}-\frac{2i^2}{n^3}+\frac{i^4}{n^5}\right)$$

$$=\pi a^2 b\left(1-\frac{2}{n^3}\left(\frac{n^3}{3}+p_2(n)\right)+\frac{1}{n^5}\left(\frac{n^5}{5}+p_4(n)\right)\right)$$

$$=\pi a^2 b\left(\frac{8}{15}+\frac{2p_2(n)}{n^3}+\frac{p_4(n)}{n^5}\right).$$

Since p_k is a polynomial of degree at most k, we can make the last two terms as small as we wish by taking n sufficiently large. This tells us that the volume of our solid can be neither larger nor smaller than $\frac{8}{15}$ths of the volume of the cylinder in which it sits, or $8\pi a^2 b/15$.

1.4

The Binomial Theorem

Fourth powers had never occurred to the Hellenistic philosophers whose mathematics was rooted in geometry, for they would suggest a fourth dimension. But by the end of the first millennium in the Middle East, in India, and in China astronomers and philosophers were using polynomials of arbitrary degree. Sometime around the year 1000, almost simultaneously within these three mathematical traditions, the binomial theorem appeared,

$$(a+b)^n = \sum_{k=0}^{n} C_k^n a^k b^{n-k},$$

where C_k^n is the $k+1$st entry of the $n+1$st row in the triangular arrangement

1

1 1

1 2 1

1 3 3 1

1 4 6 4 1

1 5 10 10 5 1

1 6 15 20 15 6 1

$\vdots$

Each entry is recognized as the sum of the two diagonally above, what today we call *Pascal's triangle*.[7] The initial purpose of this expansion was to find roots of polynomials,[8] but they would come to play many important roles in mathematics. In particular, the binomial theorem provides a means of finding sums of arbitrary positive integer powers.

The starting point for deriving a formula for the sum of kth powers is an observation of Pascal's triangle that was made many times by many different philosophers. In Figure 1.13, we see that if we start at any point along the right-hand edge and add up the terms along a southwest diagonal, then

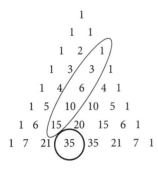

Figure 1.13. The sum of terms down a diagonal, starting from the edge, is always equal to the next term down the opposite diagonal.

wherever we choose to stop, the sum of those numbers is equal to the next number southeast of the number at which we stopped. It is not particularly difficult to see why this is so. For instance, if we take the example in the figure,

$$1 + 3 + 6 + 10 + 15 = 35,$$

$1 + 3$ is the same as summing 3 and the 1 that lies immediately to its right. From the way this triangle is constructed, $3 + 1$ equals the number directly below them and to the right of the 6. The sum of the first three terms down the diagonal is equal to the sum of the last term and the number immediately to its right. The sum of the 6 and the 4 is equal to the number immediately below them, which is the number immediately to the right of the 10 that lies along the diagonal. Wherever we choose to stop, the sum of the terms along the diagonal is equal to the last term plus the term to its right, which is the number directly below.

The earliest documented appearance of this observation occurs in an astrological text by the Spanish-Sephardic philosopher Rabbi Abraham ben Meir ibn Ezra (1090–1167). It also appears in the Chinese manuscript *Siyuan Yujian* (Jade mirror of the four origins) by Zhu Shijie, from 1303, and also in 1356 in the Indian text *Ganita Kaumudi* (Moonlight of mathematics) by Narayana Pandit (circa 1340–1400). It can be expressed as

(1.4) $$C_k^k + C_k^{k+1} + C_k^{k+2} + \cdots + C_k^{k+n-1} = C_{k+1}^{k+n}.$$

As we will see in section 1.7, Pierre de Fermat would use this insight to discover the area beneath the graph of $y = x^k$ from 0 to a for arbitrary positive integer k, the formula that today we would write as

(1.5)
$$\int_0^a x^k \, dx = \frac{1}{k+1} a^{k+1},$$

for any positive integer k.

1.5
Western Europe

The works of Euclid and Archimedes that were known to the European scientists of the sixteenth and seventeenth centuries had survived the Early Middle Ages in Constantinople, copied over the succeeding centuries by scribes who often had no understanding of what they were writing. By the eighth century, Euclid's *Elements* and Archimedes' *Measurement of a Circle* and *On the Sphere and Cylinder* had found their way from the Byzantine Empire to the courts of the Islamic caliphs who had them translated into Arabic. By the twelfth century, Latin translations of the Arabic had begun to appear in Europe. In the following centuries, Euclid was introduced into the university curriculum, but even the master's degree required attending lectures on at most the first six books, and students were seldom held responsible for anything beyond Book I.

Euclid's *Elements*, in Campanus's Latin translation of an Arabic text, was the first mathematics book of any significance to be printed. This was in Venice in 1482. It was followed in 1505 by a translation from a Greek manuscript based on a commentary on the *Elements* by Theon of Alexandria (circa 355–405 CE). Until 1808 when François Peyrard discovered an earlier version of the *Elements* in the Vatican library, the standard edition of Euclid's *Elements* was the 1572 translation by Commandino of Theon's commentary.[9]

The survival of Archimedes' work was even more tenuous. In addition to the Arabic texts, there were two Greek manuscripts, probably copied around the tenth century in Constantinople, that each contained several of his works. These are believed to have been taken to Sicily by the Normans when they conquered that kingdom in the eleventh century. At the defeat of Manfred of Sicily at the Battle of Benevento in 1266, the Archimedean manuscripts were sent to the Vatican in Rome where three years later they were translated into Latin. In 1543, Niccolò Tartaglia published Latin translations of *Measurement of a Circle, Quadrature of the Parabola, On the Equilibrium of Planes,* and Book I of *On Floating Bodies.* The following year, all of the known works of Archimedes were published in the original Greek together with a Latin translation.[10]

Federico Commandino (1509–1575) translated into Latin and then published works of many of the Greek masters: Euclid, Archimedes, Aristarchus of Samos, Hero of Alexandria, and Pappus of Alexandria. The translation into Latin and publication of Pappus's *Collection,* which would inspire both Fermat and Descartes, was completed in 1588 by his student Guidobaldo del Monte (1545–1607). Commandino and others, including Francesco Maurolico (1494–1575), expanded on Archimedes' results, especially the problem of finding centers of gravity. Maurolico determined the center of gravity of a paraboloid using inscribed and circumscribed discs of constant thickness, calculating the respective centers of gravity of these stacks of discs and showing that the distance from the apex to the center of gravity can be neither larger nor smaller than two-thirds the distance from the apex to the base.[11]

Over the following decades, the Dutch engineer Simon Stevin (1548–1620) and the Roman philosopher—and frequent correspondent of Galileo—Luca Valerio (1552–1618) applied the Archimedean techniques to determine areas, volumes, and centers of mass. As Baron[12] has pointed out, the work of Maurolico, Commandino, Stevin, and Valerio is entirely within the framework of the formal proofs received from Archimedes. In the next century, scholars searching for "quick results and simplified techniques" would begin to loosen these strictures and adopt the use of infinitesimals. By the mid-seventeenth century, these tools were sufficiently well established that Cavalieri, Torricelli, Gregory of Saint-Vincent, Fermat, Descartes, Roberval, and their successors were able to apply them to the production of many of the common formulas for solids of revolution.

The first systematic treatment of volumes of solids of revolution was the *Nova steriometria doliorum vinariorum* (New solid geometry of wine barrels) published by Johannes Kepler (1571–1630) in 1615. It included formulas for the volumes of 96 different solids formed by rotating part of a conic section about some axis. An example is the volume of an

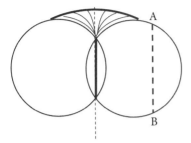

Figure 1.14. An apple formed by rotating a circle about one of its chords.

apple, formed by rotating a circle around a vertical chord of that circle (see Figure 1.14). Abandoning Archimedean rigor, Kepler established this result by considering the apple as composed of infinitely many thin cylindrical shells. We take one of the vertical chords such as AB, rotate it around the central axis, and find the surface area of this cylinder. The volume of the solid is obtained by adding up these surface areas. In practical terms, what he did was to take these cylinders, unroll each into a rectangle, and then assemble the rectangles into a solid whose volume he could compute. It is what today we refer to as the *shell method*.

There is a simpler way of computing volumes of solids of revolution that had been known to Pappus of Alexandria in the fourth century CE. In his *Collection* of the known geometric results of his time, he stated that the volume of a solid of revolution is proportional to the product of the area of the region that is rotated to form the solid and the distance from the center of gravity to the axis. Unfortunately, all that has survived is the statement of this theorem with no indication of how Pappus justified it. In 1640, Paul Guldin (1577–1643), a Swiss Jesuit trained in Rome and a regular correspondent of Kepler, published a statement and proof of this theorem in his book *De centro gravitatis*.[13]

1.6
Cavalieri and the Integral Formula

Bonaventura Cavalieri (1598–1647) was strongly influenced by Kepler. A student of Benedetto Castelli (1578–1643) who had studied with Galileo, Cavalieri began an extensive correspondence with Galileo in 1619 and discovered Kepler's *Stereometrica* around 1626. He obtained a professorship

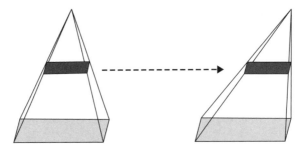

Figure 1.15. Solids with the same cross-sections have identical volumes.

in mathematics at the University of Bologna in 1629, two years after he had finished much of the work on his *Geometria indivisibilibus*. It would not be published until 1635. Galileo had been working along similar lines, and it has been suggested[14] that Cavalieri may have been waiting for Galileo to publish these results.

Cavalieri proceeded from the assumption that areas can be built up from one-dimensional lines and solids are composed of two-dimensional *indivisibles*. These were not just infinitely thin sheets. Cavalieri explicitly rejected the idea that solids could be thought of as built from three-dimensional but infinitesimally thin sheets. His starting point for computing volumes was the observation, going back to Democritus (circa 460–370 BCE), that if two solids have the same height and congruent cross-sections at each intermediate height, then they must have the same volume (Figure 1.15). Democritus had used this argument to prove that the area of any pyramid is one-third the area of the base times the height, but making the step to the assumption that the solid actually *is* a stack of these two-dimensional cross-sections went too far for many. Guldin was one of many vociferous critics.

Cavalieri's *Geometria* contains the first derivation of a formula equivalent to the integral formula for x^k. Though Cavalieri only carried this up to the integral of x^9, that was far enough that anyone could see what the general formula had to be. In explaining Cavalieri's work, it is important to recognize that this was written before the development of analytic geometry, the ability to represent a relationship such as $y = x^k$ as a graph with an area beneath it. What we today interpret as an integral Cavalieri understood as simply a sum, a sum involving lines used to build up an area.

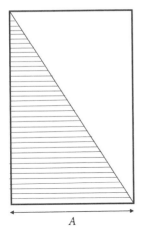

Figure 1.16. The triangular region is composed of lines
of variable length $0 \leq \ell \leq A$.

We begin with the triangular region in Figure 1.16 which shows some
of the lines that make up this triangle. Cavalieri thought of the area of this
region as the sum of the lengths of all of these lines, $\sum \ell$. The area of the
entire rectangle[15] is the sum of lines of equal length A, $\sum A$. The first step
for Cavalieri was the fact that

$$\frac{\sum \ell}{\sum A} = \frac{1}{2};$$

the area of the triangle is half the area of the rectangle.

Instead of simply summing the lengths of the lines that constitute the triangle, he now summed their squares. If we place a square of base ℓ^2 on each
line, we get a pyramid, which we have seen was long known to have volume equal to one-third of the rectangular solid formed by stacking squares
of equal size $A \times A$,

$$\frac{\sum \ell^2}{\sum A^2} = \frac{1}{3}.$$

Cavalieri now stepped into the unknown by considering the ratio of
the sum of cubes of the lines in the triangle to the sum of cubes of A. He
accomplished this using the equality

(1.6) $(x+y)^3 + (x-y)^3 = 2x^3 + 6xy^2.$

Instead of summing ℓ^3 as ℓ decreases from A to 0, he added $(A/2 + \ell)^3$ as
ℓ decreases from $A/2$ to 0 and $A/2 - \ell$ as ℓ increases from 0 to $A/2$,[16]

Figure 1.17. Pierre de Fermat.

$$\sum_{0\le\ell\le A} \ell^3 = \sum_{0\le\ell\le A/2} \left(\left(\frac{A}{2}+\ell\right)^3 + \left(\frac{A}{2}-\ell\right)^3 \right).$$

He could now use equation 1.6 and the formula he knew for $\sum \ell^2$,

$$\sum_{0\le\ell\le A} \ell^3 = \sum_{0\le\ell\le A/2} \left(2\left(\frac{A}{2}\right)^3 + 6\left(\frac{A}{2}\right)\ell^2 \right)$$

$$= \frac{1}{4} \sum_{0\le\ell\le A/2} A^3 + 3A \sum_{0\le\ell\le A/2} \ell^2$$

$$= \frac{1}{4} \sum_{0\le\ell\le A/2} A^3 + A \sum_{0\le\ell\le A/2} \left(\frac{A}{2}\right)^2$$

$$= \frac{1}{4} \sum_{0\le\ell\le A/2} A^3 + \frac{1}{4} \sum_{A/2\le\ell\le A} A^3$$

$$= \frac{1}{4} \sum_{0\le\ell\le A} A^3.$$

He proceeded up to $\sum \ell^9$, in each case using the identity

$$(x+y)^k + (x-y)^k = 2x^k + 2C_2^k x^{k-2}y^2 + 2C_4^k x^{k-4}y^4 + \cdots$$

and the formulas he had already found to show that, for $1 \le k \le 9$,

$$\frac{\sum \ell^k}{\sum A^k} = \frac{1}{k+1}.$$

If you rotate the rectangle by 90° counter-clockwise, you see that he has demonstrated that the area under the curve $y = x^k$, $0 \le x \le A$, is equal to

$$\sum_{0 \le \ell \le A} \ell^k = \frac{1}{k+1} \sum_{0 \le \ell \le A} A^k = \frac{1}{k+1} A^{k+1}.$$

Unfortunately, few people in 1635 realized what he had accomplished. Cavalieri's great work was almost unreadable.[17] What people would come to know of Cavalieri's mathematics was due to Torricelli's 1644 explanation in *Opera geometrica*. By this time, Fermat and Descartes had established algebraic geometry for graphing algebraic relationships, and they and others had found simpler routes to the integral formula.

1.7
Fermat's Integral and Torricelli's Impossible Solid

In 1636, Pierre de Fermat (1601–1665) wrote to two of his colleagues in Paris, Marin Mersenne (1588–1648) and Gilles de Roberval (1602–1675), announcing that he had discovered a general method for finding the area beneath the graph of the curve $y = x^k$ for positive integer k. Within a month, Roberval responded, stating that this result had to rely on the fact that (in modern notation)

(1.7)
$$\sum_{j=1}^{n} j^k > \frac{n^{k+1}}{k+1} > \sum_{j=1}^{n-1} j^k,$$

for all positive integers k and n. Fermat was clearly disappointed that Roberval caught on so quickly, but expressed his doubts that Roberval was able to justify this pair of inequalities.

Reconstructing Fermat's proof as best we can[18] and casting it in modern notation, the proof begins with the fact that the binomial coefficients can be written as

$$C_k^{k+j-1} = \frac{j(j+1)(j+2)\cdots(j+k-1)}{k!}.$$

We expand the numerator as a polynomial in j,

(1.8) $$C_k^{k+j-1} = \frac{1}{k!}\left(j^k + a_1 j^{k-1} + a_2 j^{k-2} + \cdots + a_k\right),$$

where the coefficients a_i are integers. Combining equation (1.4) with equation (1.8), we obtain,

(1.9)
$$\frac{1}{k!}\sum_{j=1}^{n}\left(j^k + a_1 j^{k-1} + a_2 j^{k-2} + \cdots + a_k\right) = \frac{n(n+1)(n+2)\cdots(n+k)}{(k+1)!}.$$

We can express the sum of kth powers in terms of sums of lower powers,

(1.10)
$$\sum_{j=1}^{n} j^k = \frac{k!}{(k+1)!}n(n+1)(n+2)\cdots(n+k)$$
$$- \sum_{j=1}^{n}\left(a_1 j^{k-1} + a_2 j^{k-2} + \cdots + a_k\right).$$

We use the inductive assumption[19] that the sum of mth powers from 1^m up to n^m is a polynomial in n of degree $m+1$. We have seen this to be true for $m = 1$, 2, and 3 and can assume it to be true up to $m = k-1$. Equation (1.10) is then expressed as

(1.11) $$\sum_{j=1}^{n} j^k = \frac{1}{k+1}n^{k+1} + \text{a polynomial in } n \text{ of degree at most } k.$$

To find the area under the curve $y = x^k$, we subdivide the interval from 0 to a into n subintervals of equal width, a/n (Figure 1.18). The combined area of the inscribed rectangles is

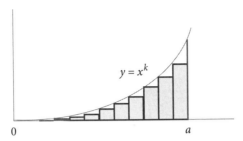

Figure 1.18. Inscribed rectangles of width a/n below the graph of $y = x^k$.

$$\sum_{j=0}^{n-1}\left(\frac{aj}{n}\right)^k\frac{a}{n} = \frac{a^{k+1}}{n^{k+1}}\sum_{j=0}^{n-1}j^k$$

$$= \frac{a^{k+1}(n-1)^{k+1}}{(k+1)n^{k+1}} + \text{a sum of terms involving}$$
$$\text{negative powers of } n.$$

This can be brought as close as we wish to $a^{k+1}/(k+1)$ by taking n sufficiently large.

The combined area of the circumscribed rectangles is

$$\sum_{j=1}^{n}\left(\frac{aj}{n}\right)^k\frac{a}{n} = \frac{a^{k+1}}{n^{k+1}}\sum_{j=1}^{n}j^k$$

$$= \frac{a^{k+1}n^{k+1}}{(k+1)n^{k+1}} + \text{a sum of terms involving negative powers of } n,$$

which also can be brought as close as we wish to $a^{k+1}/(k+1)$ by taking n sufficiently large. The area is $a^{k+1}/(k+1)$.[20]

Evangelista Torricelli (1608–1647) was another student of Castelli, earning his tuition by serving as Castelli's secretary. He began his correspondence with Galileo in 1632 and spent the last few months of Galileo's life with him, from October 1641 until January 1642. In his *Opera geometrica*, published in 1644, Torricelli embraced the language of indivisibles that Cavalieri had espoused, but he explicitly stated that his indivisibles do have "a thickness which is always equal and uniform,"[21] even though it is infinitesimal.

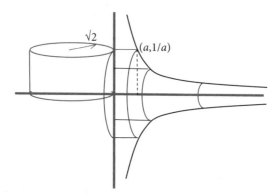

Figure 1.19. Torricelli's acute hyperbolic solid.

Torricelli is best known today—and at the time made his reputation—for the discovery of an infinitely long solid of revolution of finite volume, what he called an *acute hyperbolic solid*. This is the solid obtained by rotating about the horizontal axis the region bounded above by $y = 1/a$ for $0 \leq x \leq a$ and by $y = 1/x$, for all $x \geq a$, where a is strictly positive. Specifically, what he proved is that the volume of this solid is equal to the volume of the cylinder of radius $\sqrt{2}$ and height $1/a$ (see Figure 1.19). In other words, the volume of this infinitely long solid is the finite value $2\pi/a$.

The proof proceeds by decomposing the acute hyperbolic solid into hollow cylinders of infinitesimal thickness. The hollow cylinder at height y has radius y and circumference $2\pi y$, while the distance from the base to the hyperbolic curve is $1/y$. Every cylinder, irrespective of the value of y, has the same surface area: 2π, which is the area of a circle of radius $\sqrt{2}$. We therefore can match the volume of the acute hyperbolic solid to that of the cylinder formed by discs of radius $\sqrt{2}$ stacked from $y = 0$ to $y = 1/a$.

Torricelli shared this discovery with Cavalieri in 1641, who wrote back,

> I received your letter while in bed with fever and gout ... but in spite of my illness I enjoyed the savory fruits of your mind, since I found infinitely admirable that infinitely long hyperbolic solid which is equal to a body finite in all the three dimensions. And having spoken about it to some of my philosophy students, they agreed that it seemed truly marvelous and extraordinary that that could be.[22]

In 1643, Cavalieri communicated this result, though not the proof, to Jean-François Niceron in Paris. He passed it on to Mersenne, and soon the entire mathematical world knew about it. Torricelli published two proofs the following year as part of his *Opera geometrica*, one using the method of indivisibles as described in the previous paragraph, the other employing the classical Archimedean approach in which he demonstrated that the volume of his solid could be neither larger nor smaller than that of the cylinder of radius $\sqrt{2}$ and height $1/a$.

Torricelli's result truly shocked the mathematical establishment. He later recorded that Roberval had not believed the result when he first learned of it and had attempted to disprove it.[23] The fact that the initial proof used Cavalieri's indivisibles cast considerable doubt on their reliability, which is why Torricelli realized that he also needed to provide a justification with full Archimedean rigor.

1.8
Velocity and Distance

If accumulation were no more than a way of calculating areas, volumes, and moments, it would have provided us with an interesting set of results, but hardly the historical foundation for a major branch of mathematics. What made accumulation the powerful tool it is today was the discovery of the connection to instantaneous velocity. If we know the velocity at each point in time, then we can accumulate small changes in distance to find the total distance that has been traveled. This is not a simple or obvious idea. More than one calculus student has been mystified by the fact that we can find distances by calculating areas under curves.

Today, we take the concept of velocity of an object at a particular moment in time for granted. It confronts us every time we look at a speedometer. Yet explaining what it means requires some subtlety. The fifth century BCE philosopher Zeno of Elea described the paradox of instantaneous velocity: An arrow is always either in motion or at rest. At a single instant, it cannot be in motion, for to be in motion is to change position, and if it did change position in an instant, then that instant would have a duration and could be subdivided. Therefore, at each instant, the arrow is at rest. But if the arrow is at rest at every instant, then it is *always* at rest, and so it never moves.[24]

Aristotle answered this paradox by denying the existence of instants in time, consequently denying the existence of an instantaneous velocity. To Aristotle and his successors, this was not a great loss. The motion they studied was uniform motion, either linear or circular. There was no general treatment of velocity as the ratio of distance traveled to the time required or even as a magnitude in its own right.[25] But in the fourteenth century scholars in Oxford and Paris began to study velocity as something that has a magnitude at each instant of time and to explore what could be said when velocity is not uniform.

The first of the great European universities was established in Bologna in 1088. Others soon followed. The Greek classics, which were now being translated from Arabic, provided grist for the scholars who gathered there. They sought to understand these works. Soon they would transcend them.

Merton College in Oxford was established in 1264. Starting around 1328, a remarkable group of Mertonian scholars—Thomas Bradwardine, William Heytesbury, Richard Swineshead, and John Dumbleton—began their explorations of velocity. The first of their accomplishments was to separate kinematics, the quantitative study of motion, from dynamics, the study of the causes of motion. The idea of describing a moving object with no reference to what set that object in motion or maintained its motion was new. For the first time, scholars began to speak of velocity as a magnitude.[26]

The earliest description of instantaneous velocity can be found in William Heytesbury's 1335 manuscript, *Rules for Solving Sophisms*. He made it clear that instantaneous velocity, the velocity at a single instant of time, is not affected in any way by how far the object has moved, but is "measured by the path which *would* be described by the most rapidly moving point if, in a period of time, it were moved uniformly at the same degree of velocity."[27] This was an adequate definition that would serve for close to 500 years. It is not how we define instantaneous velocity today. Our modern definition was not fully articulated until the early nineteenth century. It is based on limits and the algebra of inequalities and can be found in section 4.2.

Heytesbury went on to consider the motion of an object that is uniformly accelerated, whose velocity increases at a constant rate. He argued that the distance traveled by an object that starts with an initial velocity and accelerates or decelerates uniformly to some final velocity is the same as the distance traveled by an object moving at the mean velocity, the velocity

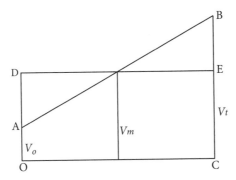

Figure 1.20. Oresme's demonstration of the Mertonian rule.

that is the average of the initial and final velocities. This is known as the
Mertonian rule. Heytesbury's argument, echoed in other proofs by his colleagues at Merton College, amounts to the observation that for a uniformly
accelerating body, the deficit in velocity over the first half of the time interval is exactly balanced by the excess over the second half. Heytesbury did
make the intriguing observation that if the acceleration is not uniform, then
nothing can be said about the distance that is traversed.[28] In fact, just a few
years later Nicole Oresme would introduce a powerful new technique to
enable the analysis of non-uniformly accelerating motion.

Oresme (1320–1392) was a scholar at the College of Navarre, founded by
Queen Joan I of Navarre in 1305, within the University of Paris. Sometime
between 1348 when he first entered Navarre as a student and 1362 when
he left to become canon at the cathedral of Rouen, Oresme wrote *On the
Configurations of Qualities*. The work deals with geometric interpretations
of qualities and includes a geometric proof of the Mertonian rule. The key
idea is to capitalize on the recognition of velocity as a magnitude and to
represent it with a line.

We indicate the time during which the object is moving by a line segment
(segment OC in Figure 1.20), each point on the line segment representing
an instant in time. Above each instant is a perpendicular line segment that
shows the intensity of the velocity at that instant. The tops of these lines
trace out a line or curve that Oresme referred to as the "line of intensity." For
an object moving at constant velocity, the line of intensity lies parallel to the
axis representing time (segment DE in Figure 1.20). For a uniformly accelerating object, the line of intensity is oblique (segment AB in Figure 1.20).

Oresme recognized the area between the axis of time and the line of intensity as the total distance traveled during the time interval. Because the area of the rectangle *OCED* is the same as the area of quadrilateral *OCBA*, the distances traveled must be the same.

We can find this argument earlier than Oresme. Giovanni di Casali in Bologna gave essentially this same geometric proof in 1346.[29] The difference lies in the clarity of Oresme's exposition and the recognition that he had discovered a general principle that did far more than prove the Mertonian rule. Oresme observed that the line of intensity might be any curve, in his words, "figures disposed in other and considerably varying ways."[30] Irrespective of how the line of intensity varies, the total distance traveled always will be represented by the area beneath this line of intensity.

Neither Oresme nor Casali made any attempt to justify that the area beneath the line of intensity represents the total distance. That would have to wait until the seventeenth century.

1.9
Isaac Beeckman

The Dutch Republic declared its independence from Spain in 1581. Over the following century, it became fertile ground for the advancement of science. One contributing factor was the freedom from the strictures of the Roman Church. When Galileo's writings were banned, he found publishers in Holland. René Descartes (1596–1650) took refuge here when his writings became too controversial for France. A second factor was the great public works, the dykes and canals that turned a swampy estuary into farmland. Simon Stevin was but one of many engineers employed in the design of locks, harbors, and mills to pump the water. In 1600, the Leiden School of Engineering was founded. Practical needs required new mathematical tools. Finally, there was money. In the seventeenth century, the Dutch Republic became one of the great centers of international trade, generating the wealth that could support scientists such as Christiaan Huygens.

Isaac Beeckman (1588–1637), a student of Stevin, is best known as friend and then rival to Descartes. It was he who introduced Descartes to Archimedean mathematics and the insights of Stevin. Beeckman and

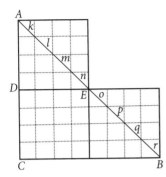

Figure 1.21. Beeckman's demonstration. The time axis is vertical, running from A to C.

Descartes met in 1618 when the young Descartes was stationed at the military school in Breda. The historian E. J. Dijksterhuis has suggested that Beeckman learned of Oresme's demonstration of the Mertonian rule from Descartes who would have studied it as part of his classical education.[31]

Beeckman provided the earliest known demonstration that an object whose velocity increases uniformly from zero will travel a distance that is proportional to the square of the elapsed time,

$$\text{distance} \propto \text{time}^2,$$

where $\propto$ is the symbol of proportionality. In a journal entry from late in 1618, Beeckman considered a falling object whose velocity increases uniformly from zero (see Figure 1.21). Note that the axes are rotated from how we normally would represent such a situation today: The time axis is vertical, and the velocity is represented by the horizontal distance from the time axis.

Beeckman began by assuming that the velocity is constant over the first interval of time, AD, represented by length DE. Over the second interval of time, velocity is represented by the constant CB. If we continue to approximate distance using two intervals with constant velocity, but double the length of the time interval and thus also the terminal velocity, the total distance is increased by a factor of four. In general, when we employ an approximation that uses two equal intervals of time with constant velocity on each interval, the total distance is proportional to the square of the elapsed time.

Figure 1.22. Galileo Galilei.

Beeckman next found a closer approximation to the true distance traveled under uniform acceleration by subdividing each interval of time into eight subintervals. The distance traveled over the time interval AC is now represented by the area of triangle ACB plus the areas of the eight small triangles denoted by k, l, m, n, o, p, q, r. Again, when the distance is approximated using eight intervals of constant velocity, the distance is proportional to the square of the elapsed time. He then argued that as we further subdivide the intervals of time, the additional area that must be added becomes progressively smaller. The sum of these additional areas "would be of no quantity when a moment of no quantity is taken." Because the proportionality between distance and the square of time does not change, the proportionality carries over to the case of a uniformly increasing velocity.[32]

Beeckman treated the velocity as constant on small intervals so that the distance is the area of a rectangle, elapsed time multiplied by velocity. He then argued that as the intervals of time become shorter, the error—the sum of the areas of these extra little pieces inserted to form rectangles—disappears. Therefore, the area between the time axis and line AB that represents the velocity under uniform acceleration is the total distance.

Beeckman never published this demonstration. Even so, it illustrates the very fertile interplay of Archimedean methods with the kinematic insights of Oresme and the Mertonian scholars. This interplay would truly flower under the influence of Galileo and his disciples.

1.10
Galileo Galilei and the Problem of Celestial Motion

Galileo Galilei (1564–1642) was born in Pisa,[33] the first child of Vicenzio Galilei, a prominent lutenist and music teacher. The family name had been Bonaiuti, but an ancestor in the early fifteenth century, Galileo Bonaiuti, had gained such fame as a physician that his descendants adopted de' Galilei as their surname.[34] In 1581, Galileo enrolled at the University of Pisa as a medical student. In the midst of his second year, he encountered the mathematician Ostilio Ricci (1540–1603), was entranced by Ricci's lectures on Euclid, and, neglecting his medical studies, turned his full attention to mathematics.

Ricci is believed to have been a student of Niccolò Tartaglia (1499 or 1500–1557), who had published many of the works of Archimedes and who played an important role in the development of algebra. It was Ricci who introduced Galileo to Archimedes as well as recent developments in the science of algebra. Galileo left the University of Pisa two years later without his medical degree and went on to hold a series of positions as a teacher of mathematics.

The focus of Galileo's work would become the heliocentric theory of planetary motion. This theory considers the earth to be one of the planets that circles the sun, in opposition to the traditional Aristotelian view that the sun and planets circle the earth. Copernicus had proposed this theory in *On the Revolutions of the Heavenly Spheres*, published in 1543, but, as a bishop, he was careful to pull his punches so that he did not come into direct conflict with his superiors in Rome. As he stated in the introduction,

> These hypotheses need not be true nor even probable. On the contrary, if they provide a calculus[35] consistent with the observations, that alone is enough. (Copernicus, 1543, pp. 3–4)

Galileo was convinced that the heliocentric theory was more than a mathematical convenience. He was certain that it described reality. But that presented him with a difficulty. If the earth makes a complete rotation once each day, then a person standing at the equator is traveling over 1000 miles per hour. Even more astounding, each year the earth must traverse almost 600 million miles as it circles the sun, requiring a speed of over 66,000 miles

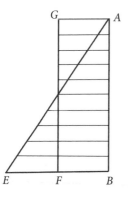

Figure 1.23. Galileo's demonstration. The time axis is vertical, running from A to B.

per hour. Yet we who stand on this spinning globe that is hurtling through space perceive no motion whatsoever. Why not?

That would become *the* great scientific question of the seventeenth century, not fully answered until the publication of Isaac Newton's *Mathematical Principles of Natural Philosophy* in 1687. Galileo's attempts to answer this question culminated in 1638 in his *Discourses and Mathematical Demonstrations Relating to Two New Sciences*. One of Galileo's greatest innovations in this book was his purely kinematic treatment of the effect of gravity on free fall. In the tradition of Oresme and the Mertonians, Galileo broke with classical tradition by ignoring the cause of gravitational attraction and focusing solely on its mathematical description. Unlike Beeckman, Galileo published his work and had a profound effect on the further development of science. The example he set of reliance on mathematics would prove so powerful that Newton, in solving this problem, would famously state, "I do not feign hypotheses,"[36] signifying that the causes and means of transmission of gravitational attraction were irrelevant to his arguments. All that really mattered was the mathematical model.

In Theorem I, under the section *Naturally Accelerated Motion* of the *Two New Sciences*, Galileo established the Mertonian rule. In Theorem II, he went on to demonstrate that for a body falling under uniform acceleration from rest, the distance traveled is proportional to the square of the time. In modern notation, if we denote the acceleration due to gravity by g, then at time t the object will have reached a velocity equal to gt. The area of the resulting triangle is $\frac{1}{2}t \times gt = \frac{1}{2}gt^2$.

Figure 1.24. Sir Isaac Newton.

In proving the Mertonian rule, Galileo drew a figure very similar to Beeckman's (see Figure 1.23). He then argued,

> Since each and every point on line *AB* corresponds to each and every instant of time *AB* and since the parallel lines drawn from these points and included in triangle *AEB* represent the growing degrees of the increasing velocity, while the parallel lines contained within the rectangle represent in the same way just as many degrees of nonincreasing but uniform velocity, it appears that there are assumed to be just as many moments of velocity in the accelerated motion represented by the growing parallel lines of triangle *AEB* as there are in the uniform motion represented by the parallel lines of *GB*. (Galilei, 1638, Naturally Accelerated Motion, Theorem I)

Galileo treated the triangle and rectangle as composed of infinitely many of these parallel lines, these "moments of velocity," each representing the distance traveled over an infinitesimal moment of time. This is reminiscent of Archimedes' derivation of the formula for the volume of a sphere explained in section 1.1 and Kepler's approach to solids of revolution. Although Galileo never lost his respect for full Archimedean rigor, he probably encouraged Cavalieri and Torricelli in their use of indivisibles.

1.11
Solving the Problem of Celestial Motion

Galileo was convinced that the answer to the problem of celestial motion, drawing on an understanding of inertia and gravitational acceleration, would come from mathematics. His derivation of the formula for the distance traveled by a falling body was part of his search for a mathematical explanation. He was prescient in recognizing the ingredients of the ultimate solution, but died before it came together. It was this Galilean agenda that motivated much of the succeeding work on accumulation, tangents, and rates. The ultimate solution would be provided by Isaac Newton (1643–1727), born almost exactly one year after Galileo's death.[37]

Newton graduated from Cambridge in 1665. Because of the plague then rampant across the cities and towns of England, he spent the next two years home in Lincolnshire. It was here that he solved the problems of heavenly motion. The philosophers who followed Galileo, especially Christiaan Huygens from the Netherlands, had developed and refined the notion of inertia, recognizing that there is no way of distinguishing between an object at rest and an object traveling in a straight line at a constant velocity. The only difference is the frame of reference. If the motion of the earth were simply linear, we could not tell that we were traveling at 107,000 km per hour relative to a frame in which the sun is stationary.

Of course, our motion is not purely linear. First of all, we rotate around the earth's axis. A speed of 1600 km per hour translates into a little less than two centimeters per second away from a straight line. If a person standing at the equator were to maintain their linear velocity, then at the end of one second the earth would have turned through 15 seconds of arc and that person would see the earth drop by $6{,}371{,}000(1 - \cos 15'') = 0.017$ m ($6{,}371{,}000$ m being the radius of the earth). The reason that we do not float off the earth is because of gravitational acceleration, which changes our velocity by adding a component directed toward the center of the earth of 9.8 meters per second for each second. This more than compensates for the 0.017 meters per second carrying us away from the earth. And, of course, the adjustments do not occur in one second intervals, but continuously. The net effect is that we are constantly falling toward the center of the earth, an effect that is normally felt as the weight of gravity but which readily manifests as downward velocity if we should step off a cliff. The same explanation accounts for

Figure 1.25. Christiaan Huygens.

our lack of awareness of the tremendous speed at which we circle the sun. The radius of the circle is much larger, about 150,000,000,000 m, but the arc is much smaller, about 0.04 seconds of arc per second, requiring an adjustment of only 0.003 meters per second.

But that raises another question. We know about gravity on earth. To the ancient Greek philosophers, it was simply the tendency of all solids and liquids to move toward earth's center. Galileo had demonstrated that gravity needs to be thought of in terms of an acceleration toward the center of the earth. Keeping the earth in an orbit around the sun would require an acceleration or gravitational attraction toward the sun. Could such a thing exist?

The story of Newton and the falling apple appears to be legitimate. In 1726, Newton recounted this to William Stukeley, who would go on to record it in 1752. Newton said nothing about the apple hitting him on the head, but watching the apple's fall did get him thinking about gravitational acceleration. If gravity is not a purely terrestrial phenomenon, then it might explain what keeps the moon in its orbit around the earth.

Newton realized that the gravitational acceleration from the earth acting on the moon should be less than the acceleration we on the surface of the earth experience. How much less? The starting point was a result discovered by Huygens.

Christiaan Huygens (1629–1695) entered the University of Leiden in 1645 to study law, but he also worked with Frans van Schooten on problems of determining centers of mass. When Van Schooten published his commentary on Descartes' *La Géometrie* in 1649, he included an example contributed by Huygens.[38] Huygens continued his studies at the Athenaeum in Breda. While there, he corresponded with Fr. Mersenne, a friend of his father and, as we will see, a key player in the development of differential calculus. In the early 1650s, Huygens published results on the computation of π and worked on the problem of rectification of curves (finding their length). It was on a trip to Paris in 1655 that he learned of the work of Fermat and Pascal on probabilities, and he published his own book on the subject, *On the Computation of Games of Chance*, in 1657.

Huygens gained his greatest fame for his work in astronomy and mechanics. Grinding his own lenses and constructing his own telescopes, he was the first to spot Saturn's moon Titan. A year later, in 1656, after constructing a 25-foot telescope, he determined that what Galileo had described as the "ears" of Saturn were, in fact, rings around the planet. During the late 1650s he designed the first working pendulum clock. The idea of regulating a clock by means of a pendulum went back at least to da Vinci. Both Galileo and Torricelli had attempted to design such a clock, but Huygens was the first to solve the problems of the escapement and so enable the construction of the first working model. His research into the mechanics of the pendulum clock was published in 1673 as *Horologium Oscillatorium*, a work that was widely praised and that greatly influenced Newton's own work on mechanics.

To explain Huygens's insight into circular motion, we begin by imagining a rock tied to the end of a string and then swung in a circle. We feel a tug on the string that is often referred to as centrifugal force. It appears that the swirling rock is pulling on our hand. In fact, what we feel is the force, or acceleration, we must exert to keep pulling the rock back into circular motion and preventing it from flying off in a straight line. Huygens had studied this acceleration and demonstrated that it is proportional to the square of the speed of the rock and inversely proportional to the radius of the circle, or

$$a = c\frac{v^2}{r},$$

where a is the acceleration, v is the speed, r is the radius, and c is a constant. Doubling the speed increases the needed acceleration by a factor of four. Doubling the radius cuts the acceleration in half. Because speed is distance over time, we rewrite v as $2\pi r/t$, where t is the period, i.e., the time it takes to complete one revolution,

$$a = (4\pi^2 c)\frac{r}{t^2}.$$

Kepler had observed a curious phenomenon of all of the planetary orbits, that the square of the period is proportional to the cube of the distance from the sun. If we measure the period in years (earth's period) and distances in astronomical units (the distance from the earth to the sun), then we get exact equality:

$$t^2 = r^3.$$

Putting this together, Newton realized that the acceleration required for planetary orbits, and presumably all orbits, satisfies

$$a = (4\pi^2 c)\frac{1}{r^2}.$$

Acceleration due to gravity is inversely proportional to the square of the distance.

Now the radius of the moon's orbit is about 60 times the radius of the earth, so the earth's gravitational effect on the moon is about 1/3600 times its value on the surface of the earth, or about 0.00272 meters per second per second. The moon travels 2.4 billion meters in a period of 27.3 days (a lunar month is slightly longer because the earth has moved, lengthening the time between phases), or about a thousand meters per second. In one second, it moves through 0.55 seconds of arc, which means that it needs to drop by

$$385{,}000{,}000(1 - \cos 0.55'') = 0.00137 \text{ m}.$$

Given that the downward acceleration is 0.00272 meters per second per second, Galileo's formula for distance implies that the moon will fall half that, or 0.00136 meters in this second. Except for the small difference that arises from round-off, the moon's orbit can be entirely explained by the moon falling toward the earth by 0.00137 meters as it moves forward by approximately 1025 meters in each second.

1.12
Kepler's Second Law

This insight into the effects of inertia and gravitational acceleration solved the problem of why we are unaware of the earth's tremendous speed. Newton realized it does much more. We can accumulate these small changes in velocity to see how velocity changes. Knowing the velocity at each point in time, we can now accumulate the small changes in position to determine how the position changes. Because the direction and magnitude of acceleration due to gravity is entirely determined by the position of the orbiting body, position determines acceleration. We are now in a "virtuous cycle": Acceleration determines velocity which determines position which determines acceleration. If we know the initial velocity and position of an orbiting body, then, in the absence of any additional forces, the path it will take is uniquely determined. Nowhere is this clearer than in Newton's proof of Kepler's observation that orbiting planets sweep out equal area in equal time (Figure 1.26), thus speeding up when closest to the sun and slowing down when it is farthest. It is Proposition I in his *Mathematical Principles of Natural Philosophy*:

> The areas which bodies made to move in orbits describe by radii drawn to an unmoving center of forces lie in unmoving planes and are proportional to the times.

Begin with an object that travels along the straight line *l* at a constant velocity, moving the same distance in each small interval of time (Figure 1.27).

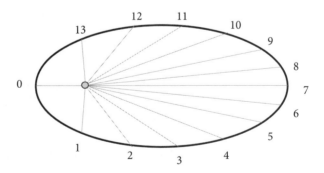

Figure 1.26. Sweeping out equal area in equal time.

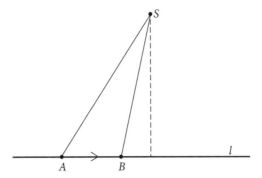

Figure 1.27. An object moving at constant velocity sweeps out equal area in equal time.

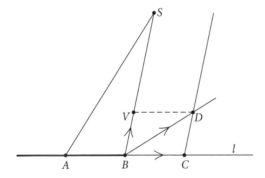

Figure 1.28. If the only acceleration is radial, then the object still sweeps out equal area in equal time.

The area swept out in this interval of time is the area of the triangle ABS, which equals the distance from S to the line l (the height of the triangle) times $\overline{AB}$ (the length of the base). Since the height and length of the base do not change as we move along this line at a constant velocity, the area swept out is the same over each interval of time.

We now introduce a small change in velocity directed toward S and represented by the vector from B to V (Figure 1.28). Adding this to the original velocity means that in the next interval of time, our object moves from B to D, where the line through C and D is parallel to that through B and V. The area swept out in this second interval of time is the area of the triangle BDS. If we compare it to the triangle BCS, we see that they share a base, namely the line segment from B to S, and the same height, the perpendicular distance between the line through B and V and the line

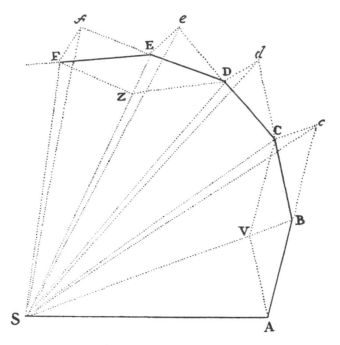

Figure 1.29. Newton's representation of the proof of Kepler's law.

through C and D. Because they have the same area, the area swept out in the second interval of time equals the area swept out in the previous interval.

As the object continues to move, constantly adjusting its velocity by adding components directed toward S, it will continue to sweep out equal areas over equal intervals of time (Figure 1.29). As the intervals of time get smaller, the path approaches a smooth curve.

1.13
Newton's *Principia*

One possible date for the birth of calculus is 1666 when Newton discovered the connection between accumulation (integration) and ratios of change (differentiation). The realization of the importance of the tools that today we identify with calculus emerged over the following decades. One of the strongest indicators of their power came, paradoxically, from the publication of Newton's *Principia* in 1687. It is paradoxical because the *Principia* contains almost nothing that today we would recognize as calculus. It is entirely couched in the language of Euclidean geometry. But it deals with problems of accumulation, ratios of change, and rates of change that today we would state in terms of calculus.[39]

The creation began in 1684 when Edmond Halley, Christopher Wren, and Robert Hooke, realizing that planetary orbits would be completely determined by the fact that gravitational attraction is inversely proportional to the square of the distance, wondered why that implied an elliptical orbit with the sun at one focus. The mathematics of it stumped all three.

That summer, Halley visited Newton in Cambridge and posed this problem. Newton replied that he had already worked this out, but that he could not find his original derivation. Halley encouraged him to write it up. Later that year Newton produced *De Motu Corporum* (On the motion of bodies), a tract that accomplished this task. But Newton was not satisfied with it. The true situation is much more complicated. Not only does the sun pull on the earth, the earth also pulls on the sun. In fact, every planet exerts a gravitational attraction on every other planet. The effect of Jupiter on Earth is minimal, but the motion of the moon around the earth is very much affected by the sun.

Furthermore, the kind of analysis done to prove Kepler's second law assumes that the orbiting planet can be represented by a single point in space. What happens when we take into account the fact that it is a solid sphere? In addition, he realized that he needed to explain how these elliptical orbits translated into the paths of the planets across the heavens as we see them from earth, and he needed to be able to compare the results of his mathematical models with actual astronomical observations. It was not just a question of elaborating his models. He also needed to explain and justify his assumptions, explaining what he meant by mass, force, and inertia and justifying the tools of accumulation and ratios of change that he needed to accomplish this work. The entire tome came to 510 pages (Figure 1.30).

There is nothing in this book that looks at all like calculus as we think of it today. There are no explicit derivatives or integrals. But the essential ideas of calculus, what he calls "the method of first and ultimate ratios" and which amount to accumulations and ratios of change,

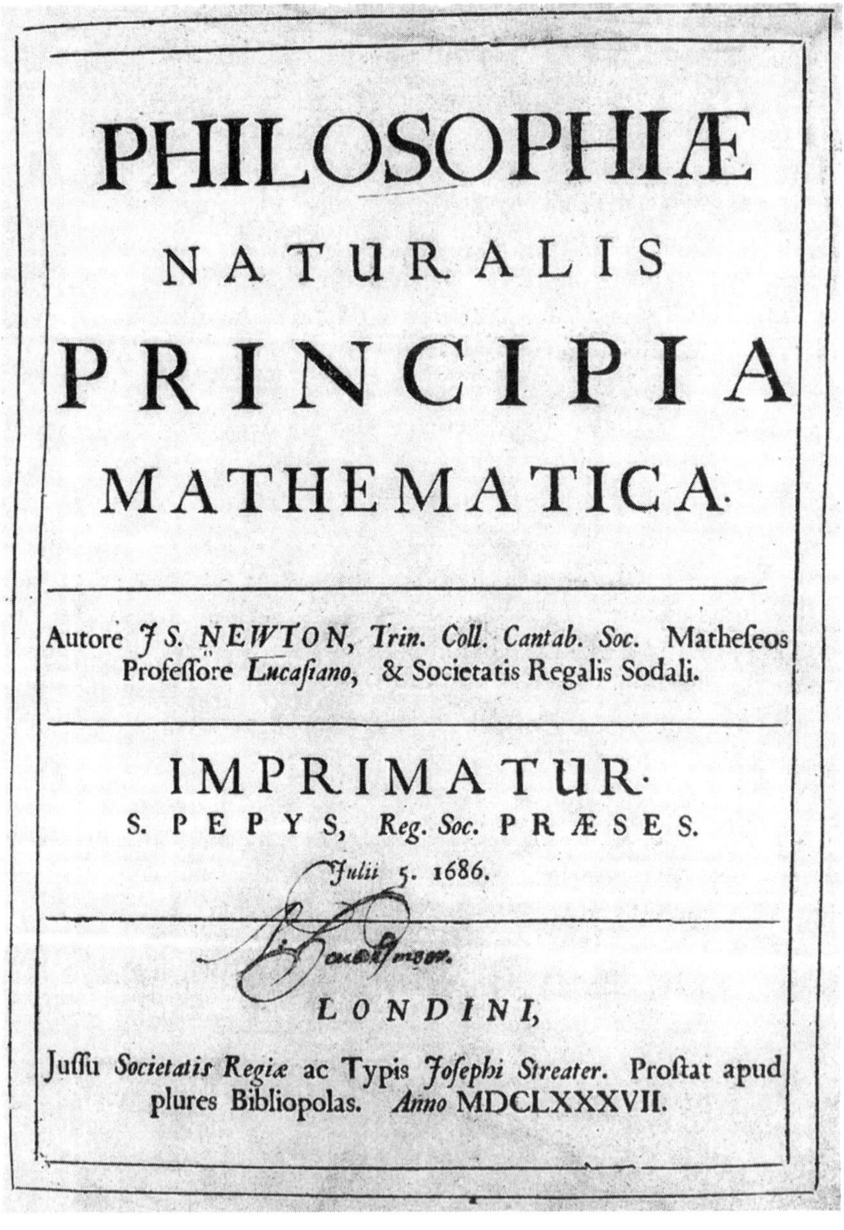

PHILOSOPHIÆ
NATURALIS
PRINCIPIA
MATHEMATICA.

Autore *J S. NEWTON, Trin. Coll. Cantab. Soc.* Mathefeos Profeffore *Lucafiano*, & Societatis Regalis Sodali.

IMPRIMATUR.
S. PEPYS, *Reg. Soc.* PRÆSES.
Julii 5. 1686.

LONDINI,

Juffu *Societatis Regiæ* ac Typis *Jofephi Streater*. Proftat apud plures Bibliopolas. *Anno* MDCLXXXVII.

Figure 1.30. Title page to first edition of Newton's *Principia*.

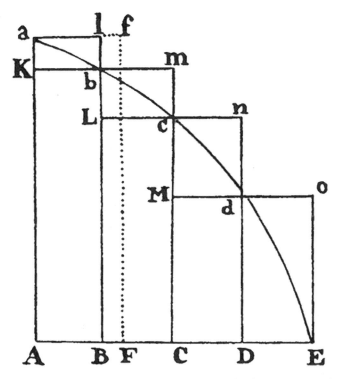

Figure 1.31. Newton's illustration for the proofs of Lemmas 2 and 3.

undergird every one of the 192 propositions that he proves within its covers. He begins Book I with eleven lemmas that establish the foundations of calculus.

Lemma 1 defines what today we would call the limit (see section 4.1 for further details). Lemmas 2 and 3 show that any area, and by extension any accumulation problem, can be evaluated by taking circumscribed and inscribed rectangles over sufficiently small intervals. Figure 1.31 displays Newton's illustration. In Lemma 2 he assumes that the bases of the rectangles are equal, that $\overline{AB} = \overline{BC} = \overline{CD} = \overline{DE}$. He observes that if we take the small rectangles *Kbla*, *Lcmb*, *Mdnc*, and *DEod* that correspond to the difference between the areas of the circumscribed and the inscribed rectangles, they all can be slid above the interval *AB*. The sum of the areas of these difference rectangles is exactly the area of the rectangle *ABla*. As we take more and narrower intervals, the differences will fit inside a rectangle of the same height, but of diminishing width, so that its area can be made as small as we wish. Thus the areas given by the circumscribed and inscribed rectangles "approach so close to one another that their difference is less than any given quantity."

In Lemma 3, he observes that the lengths of the bases do not need to be equal. If one is wider than the others, say the base of rectangle *AFfa*, then the differences still fit inside a rectangle whose area can be made as small as we wish by taking the longest base to be sufficiently small.

To prove that the orbits of the planets are ellipses with the sun at one focus, Newton relies on the fact that once the initial position and velocity are known, the path is uniquely determined. He then proves that an elliptical orbit with the gravitational attraction emanating from one focus satisfies the property that the acceleration (change in velocity) is inversely proportional to the square of the distance. By the uniqueness of the solution[40] this must be the path that planets follow.

Now we are looking at the inverse problem. Rather than using information about the acceleration to accumulate changes in velocity and position, we are using information about the position and velocity to determine the acceleration. We are now entering what would come to be known as the differential calculus or the study of ratios of change.

Chapter 2

RATIOS OF CHANGE

It is common to introduce the derivative as the slope of a tangent line. Such an approach can create pedagogical difficulties. Philosophers and astronomers worked with ratios of change long before anyone computed slopes. Students who do not understand slope as a ratio of changes can fail to appreciate the full power of this concept.

It is possible to trace the origins of differential calculus to problems of interpolation. Given a functional relationship, say pairing numbers with their squares, it is often the case that the correspondence is easy:

$$1 \to 1, \quad 2 \to 4, \quad 3 \to 9, \quad 4 \to 16, \quad 5 \to 25, \ldots.$$

But what if we want to know what corresponds to 2.5? The value 2.5 lies halfway between 2 and 3; a first approximation might be to look halfway between 4 and 9, which is 6.5. This is simple linear interpolation, assuming the ratio of changes stays constant. As we know, it produces an answer that is close but not exact. The true value is 6.25. The ratio of the change in the input to the change in the output does not stay constant. As the input rises from 2 to 2.5, the output only increases by 2.25. As the input goes on to increase from 2.5 to 3, the output must rise faster, by 2.75.

This has been a simple example. Squaring numbers is not hard, and we do not need a sophisticated understanding of how these ratios of change are changing. But when, in the middle of the first millennium of the Common Era, astronomers in South Asia needed to find intermediate values using tables of the sine function, they realized that they required more than linear interpolation.

The connection to problems of interpolation illuminates a key aspect of differential calculus: It is a tool for understanding how change in one quantity is reflected in the variation of another to which it is linked. Problems of accumulation arose from geometric considerations and can be understood outside the context of functions. Ratios of change only make sense in the presence of functional relationships.

This chapter begins with the earliest derivatives, those of the sine and cosine that arose from the work on trigonometry in India in the middle of the first millennium of the Common Era and through John Napier's invention of the natural logarithm in the opening years of the seventeenth century. Slopes of tangent lines and derivatives of polynomials would not arrive until the tools of algebra and algebraic geometry had been well established. In preparation for this part of the story, we will travel rapidly through a brief history of algebra. During the seventeenth century, dozens of philosophers worked to develop the tools we today recognize as calculus, finally culminating in Newton's discovery of the Fundamental Theorem of Integral Calculus connecting problems of accumulation to an understanding of ratios of change.

As derivatives came to be understood as rates of change, they opened the door to modeling the dynamic interplay behind many physical phenomena. We have seen one example of this in Newton's exploration of celestial mechanics. Beginning in the eighteenth century, this application of calculus became a powerful tool enabling major scientific advances. It is a significant flaw in our educational system that so few students of calculus ever get the chance to fully appreciate its power. For this reason, this chapter concludes with an indication of the role of differential equations in fluid dynamics, vibrating strings, and electricity and magnetism.

2.1
Interpolation

The first true functional relationship was that between the *chord* and the arc of a circle, introduced by Hipparchus of Rhodes in the second century BCE in his astronomical work. The chord is simply the length of the line segment connecting the two ends of the arc of a circle (Figure 2.1). Degrees had been introduced by Mesopotamian astronomers not as angles but as a

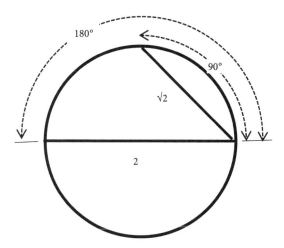

Figure 2.1. The chords corresponding to arc lengths 90° and 180° on a circle of radius 1.

measure of the arc of a full circle. Specifically, they indicate a fraction of the full circular arc whose length was 360°. Although the year is not 360 days, one degree in the sun's progression is quite close to a single day.

Given a circle of radius 1, as the arc length increases from 0 to 90° to 180°, the chord length will increase from 0 to $\sqrt{2}$ to 2 (Figure 2.1). We say that the arc and chord lengths *co-vary*. There are certain arc lengths for which Hellenistic astronomers knew how to determine exact values of the chord lengths, including

$$60° \to 1 \quad \text{and} \quad 36° \to \frac{\sqrt{5}-1}{2}.$$

Intermediate values would require interpolation.

In the early centuries of the Common Era, astronomers in South Asia learned of the Hellenistic work in astronomy and developed it further. They are the ones responsible for using the half-chord, what today we call the "sine," instead of the chord (Figure 2.2). The story of how this half-chord came to be called a sine is worth retelling. The Sanskrit for "chord" is *jya* or *jiva*, which later Arab scholars transcribed as *jyba*. But the "a" in *jyba* is written with a diacritical mark that is often omitted. While *jyba* is not a word that is otherwise used in Arabic, *jayb*, which uses the same "jyb,"

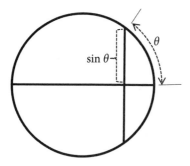

Figure 2.2. The sine of θ is half the chord of 2θ.

means "pocket." By the time the word was translated into Latin, it was understood to mean "pocket," for which the Latin word is *sinus*, hence the English word sine.

Certain values of the sine can be found exactly:

$$\sin 60° = \frac{\sqrt{3}}{2},$$

$$\sin 45° = \frac{\sqrt{2}}{2},$$

$$\sin 30° = \frac{1}{2},$$

$$\sin 36° = \frac{\sqrt{10 - 2\sqrt{5}}}{4}.$$

The last of these is taken from geometric theorems in Euclid's *Elements*. It is possible to find the exact value of the sine of any angle from 0° to 90° that is a multiple of 3° using the sum-of-angle, difference-of-angle, and half-angle formulas,[1]

$$\sin(\alpha + \beta) = \sin\alpha \, \cos\beta + \cos\alpha \, \sin\beta,$$

$$\sin(\alpha - \beta) = \sin\alpha \, \cos\beta - \cos\alpha \, \sin\beta,$$

$$\sin(\alpha/2) = \sqrt{\frac{1 - \cos\alpha}{2}}, \quad 0° \leq \alpha \leq 360°,$$

together with the fact that $\cos\alpha = \sqrt{1-\sin^2\alpha}$, $0° \leq \alpha \leq 90°$. For example,

$$\sin 39° = \sin(36° + 3°) = \sin 36° \cos 3° + \cos 36° \sin 3°,$$

$$\sin 3° = \sqrt{\frac{1 - \cos 6°}{2}},$$

$$\sin 6° = \sin(36° - 30°) = \sin 36° \cos 30° - \cos 36° \sin 30°.$$

With some work, it is possible to put this together to get an exact value for $\sin 39°$,

$$\sin 39° = \frac{(1 - \sqrt{3})\sqrt{10 - 2\sqrt{5}} + (1 + \sqrt{3})(1 + \sqrt{5})}{8\sqrt{2}}.$$

To find all of the values that they would need, Indian astronomers interpolated between the values of the sine at multiples of $3°$. Let us take as an example the sine of $37°$ when we know that, to six significant digits,

$$\sin(36°) = 0.587785 \quad \text{and} \quad \sin(39°) = 0.629320.$$

Because 37 is one-third of the way from 36 to 39, it makes sense to estimate the desired value by

$$\sin(37°) \approx 0.587785 + \frac{1}{3} \times (0.62932 - 0.587785) = 0.601630.$$

That is not bad, but in fact the true value is equal to 0.601815 (to six significant digits). What we need is a means of tackling the change in the length of the sine, or half-chord, as a proportion of the change in the arc length. We can begin by using the sine at other multiples of $3°$. The sine of $33°$ is 0.544639. From $33°$ to $36°$, the value of the sine increases by $0.587785 - 0.544639 = 0.043146$. From $36°$ to $39°$, the sine increases by $0.629320 - 0.587785 = 0.041535$. The ratio of the change in the sine to the change in the arc length is decreasing, leading us to expect that simple linear interpolation should underestimate the true value (see Figure 2.3).

Around the end of the fifth century CE, the Indian astronomer Aryabhata (476–550) realized that he could use the sum-of-angles formula for the sine

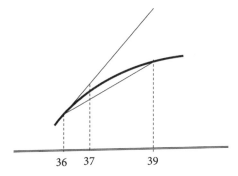

Figure 2.3. Between 0° and 90°, the sine function is increasing and concave down. Linear interpolation underestimates the true value at 37°. Using the tangent line overestimates the true value. (The concavity of the sine function has been exaggerated to illustrate this point.)

to get a better estimate of the relationship between the change in the arc length and the change in the sine. Letting $\Delta(\sin\theta)$ denote the change in $\sin\theta$ and $\Delta\theta$ the change in θ, we can write

$$\Delta(\sin\theta) = \sin(\theta + \Delta\theta) - \sin(\theta)$$
$$= \sin(\theta)\cos(\Delta\theta) + \cos(\theta)\sin(\Delta\theta) - \sin(\theta)$$
$$= (\cos(\Delta\theta) - 1)\sin(\theta) + \cos(\theta)\sin(\Delta\theta).$$

If we have a very small value for $\Delta\theta$, then $\cos(\Delta\theta)$ will be very close to 1, and the first term will be negligible,

(2.1) $$\Delta(\sin\theta) \approx \cos(\theta)\sin(\Delta\theta).$$

Indian astronomers had one big advantage over the Greeks: They were measuring both the arc length that describes the angle and the sine (or half-chord) in the same units. One degree is 1/360 of the circumference. For a circle of radius R, one degree represents the distance $2\pi R/360$. Most Indian astronomers used a circumference of 21,600 ($= 360 \times 60$, the number of minutes in a full circle) and a radius of 3438 (the value of $21,600/2\pi$ rounded to the nearest integer).

As long as we are using the same units to measure both the arc length and the half-chord or sine, it is apparent that for very small arc lengths, the two are very close, $\sin(\Delta\theta) \approx \Delta\theta$. Combining this with equation (2.1) yields

$$\Delta(\sin\theta) \approx \cos(\theta)\Delta\theta.$$

Thus, the constant of proportionality, $\Delta(\sin\theta)/\Delta\theta$, can be approximated by the cosine function at θ.

Let us go back to our problem of finding the sine of 37°, but convert the angle or arc length into the same unit of length as the radius, which we take to be 1. The value of $\Delta\theta$ is

$$\Delta\theta = 1° = 1° \times \frac{2\pi}{360°} \approx 0.0174533.$$

Because we know the exact value of $\sin(36°)$, we also know the exact value of

$$\cos(36°) = \sqrt{1 - \sin^2(36°)} \approx 0.809017.$$

Now we put it all together:

$$\sin(37°) = \sin(36°) + \Delta(\sin\theta)$$

$$\approx \sin(36°) + \cos(36°) \cdot \Delta\theta$$

$$\approx 0.587785 + 0.809017 \cdot 0.0174533 = 0.601905.$$

The improvement is not great. Simple linear interpolation underestimated the true value of $\sin(37°)$ by 0.000185. Aryabhata's improved method overestimated by 0.000090. We have cut the error in half (Figure 2.3). The difference is between finding values of the sine function near 36° by using the line that connects the graph of $y = \sin\theta$ at $\theta = 36°$ and at $\theta = 39°$ versus using the tangent line at $\theta = 36°$.

Today we write Aryabhata's relationship as

$$\lim_{\Delta\theta \to 0} \frac{\Delta(\sin\theta)}{\Delta\theta} = \cos(\theta).$$

What Aryabhata presented in his astronomical work known as the *Aryabhatiya*, written in 499, is the relationship that today we recognize as the rule for the derivative of the sine. One could claim that the first function to be differentiated was the sine, it happened in India, and it occurred well over a thousand years before Newton or Leibniz were born.

This history illuminates two important lessons. The first is the importance of measuring arc length and sine in the same units. It was Leonhard Euler in the eighteenth century who finally standardized the practice of defining the trigonometric functions in terms of a circle of radius one so that one full circumference would be 2π, leading to what today we refer to as radian measure of angles. In that sense, radian measure is relatively recent, less than 300 years old. But in the sense in which radian measure is simply a way of employing the same units for the arc length and the radius, it is the approach to measuring angles that has been used for well over 1500 years.

The second lesson is that the core idea behind the derivative, a ratio of the changes in two linked variables, does not make its first appearance as a slope or even as a rate of change, but as a tool for interpolation, enabling us to relate the changes in the input and output variables and understand how that ratio changes. This is why I refer to it as *ratios of change*.

Indian astronomers never studied derivatives as such, but they became very adept at using the fact that $\cos\theta$ can be substituted for $\Delta(\sin\theta)/\Delta\theta$ when $\Delta\theta$ is very small. Bhaskara II (1114–1185) used this knowledge to construct quadratic polynomial approximations to the sine and cosine.

In the fourteenth and fifteenth centuries, Madhava of Kerala (circa 1350–1425) and his followers combined their insight into these ratios of change with knowledge of infinite geometric series and formulas for sums of integral powers to derive infinite series expansions for the sine and cosine and to demonstrate a sequence of very close approximations to π,[2]

$$(2.2) \qquad \pi \approx \frac{4}{1} - \frac{4}{3} + \frac{4}{5} - \cdots + (-1)^{n-1}\frac{4}{2n-1} + (-1)^n\frac{4}{(2n)^2+1}.$$

Although there is no indication that any of the Indian knowledge of ratios of change or infinite series found its way to the West, problems of interpolation and approximation became important to Chinese, Islamic, and, eventually, European philosophers. They would continue to drive the development of the fundamental ideas of calculus.

2.2
Napier and the Natural Logarithm

While we have seen that European philosophers began to study velocities in the fourteenth century, these were not considered to be rates of change, that is, ratios of distance over time. Instantaneous velocity was simply a magnitude, not a ratio. This began to change in the late sixteenth century.

John Napier, Laird of Merchiston (1550–1617), was a Scottish landowner with a passion for astronomy and mathematics. One of the great challenges for astronomers in the late 1500s was the need to carry out complex calculations to high precision. Eight- and even ten-digit accuracy was often required. Just multiplying together two ten-digit numbers is a daunting task filled with many opportunities for error. Napier came upon a brilliant idea to simplify this, giving rise to today's logarithmic function. Before explaining what Napier accomplished, it is important to explain the role of the logarithm in facilitating multiplication.

If we have a table of the powers of 2:

$$2^1 = 2, \quad 2^2 = 4, \quad 2^3 = 8, \ 2^4 = 16, \quad 2^5 = 32, \quad 2^6 = 64,$$
$$2^7 = 128, \quad 2^8 = 256, \quad 2^9 = 512, \quad 2^{10} = 1024, \quad 2^{11} = 2048,$$

and if we want to multiply 8×128, we convert each to a power of 2: $8 = 2^3$ and $128 = 2^7$. We can multiply these powers of 2 by adding their exponents:

$$8 \times 128 = 2^3 \times 2^7 = 2^{3+7} = 2^{10}.$$

Going back to our table, we see that $2^{10} = 1024$ and immediately get the answer. Instead of doing any multiplication, we only had to add, a much easier operation.

Of course, the table given above is very limited and would not be much help in practice. But let us assume that we have a table that gives us the power of ten that yields each number from 1.0000001 up to 10.0000000, in increments of 0.0000001. Let us assume that we want to multiply 3.67059 by 7.20954.

With an appropriate table, we can look up the desired exponents and observe that

Figure 2.4. John Napier.

$$3.67059 = 10^{0.56473588}, \quad \text{and} \quad 7.20954 = 10^{0.85790756}.$$

Armed with these values, we find the product by adding the exponents and then looking up the result in our table:

$$3.67059 \times 7.20954 = 10^{0.56473588} \times 10^{0.85790756}$$
$$= 10^{0.56473588+0.85790756}$$
$$= 10^{1.42264344}$$
$$= 10 \times 10^{0.42264344}$$
$$= 10 \times 2.6463266$$
$$= 26.463266,$$

which is extremely close to the true product: 26.4632654 ... and accurate to 10^{-5}. All we had to do was add two eight-digit numbers.

I do need to interject that neither Napier nor any of his contemporaries would ever have written anything like $10^{0.56473588}$. Exponents could only be positive integers, providing a shorthand for multiplication of a number by itself a certain number of times. The idea of fractional and negative exponents would have to wait until John Wallis in 1656. The concept of an exponential *function*, where any real number can be inserted into the exponent to produce a corresponding positive real number, would not appear until the eighteenth century. However, it is pedagogically useful to speak of these numbers as exponents.

Napier invented the term *logarithm*. He never explained how he chose this name. The *Oxford English Dictionary* states that it arises from *logos* meaning ratio and *arithmos* meaning number. These are ratio numbers. The problem with this explanation is that translating *logos* as "ratio" is a stretch. Its usual meaning is "word," "speech," "discourse," or "reason." My personal interpretation is that Napier's term arose from the fact that Greek philosophers understood mathematics to be composed of two distinct components, *logistiki* or the art of calculation (whence our word "logistics") and *arithmetiki* or the science of numbers. Logarithms are constructs that draw on the science of numbers to facilitate calculation.

Napier's actual starting point was to construct a relationship between two sets of numbers so that ratios in one set would be reflected in differences in the other. If we denote the relationship with the function notation NapLog, then it must satisfy the relationship

$$(2.3) \qquad \frac{a}{b} = \frac{c}{d} \iff \text{NapLog } a - \text{NapLog } b = \text{NapLog } c - \text{NapLog } d.$$

The modern logarithm also satisfies this relationship. It is the key to turning problems of division or multiplication into problems of subtraction or addition. But Napier's logarithm looks strange to our eyes because he then defines $\text{NapLog } 10^7 = 0$ and, in general,

$$(2.4) \qquad \qquad \text{NapLog } 10^7 r^n = n,$$

for some suitably chosen ratio $0 < r < 1$.[3]

Today we use the function notation $\log(x)$ to denote the transformation from a multiplicative to the additive system. The defining characteristic of this function is that it converts a product into a sum,

$$(2.5) \qquad \qquad \log(xy) = \log(x) + \log(y).$$

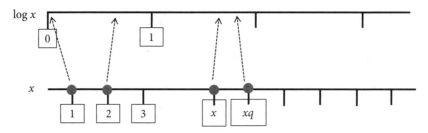

Figure 2.5. The representation in terms of modern logarithms of Napier's two number lines.

An immediate consequence is that log(1) must be 0 because

$$\log x = \log(x \cdot 1) = \log(x) + \log(1), \text{ and so } \log 1 = 0.$$

In terms of modern logarithms, Napier's logarithm is given by

$$\text{NapLog } x = \log_r \left(x 10^{-7} \right) = \log_r x - 7 \log_r 10,$$

where r is the value in equation (2.4), the *base* of the logarithm. Napier needed to choose a value for r.

To clarify the argument for modern readers, I will translate it into the language of logarithms that satisfy equation (2.5).[4] We will also take as our base $r^{-1} > 1$ so that the value of the logarithm increases as x increases. In a move that anticipated much of calculus, Napier explored the rate at which his logarithm changes as x changes. In particular, he considered two number lines (Figure 2.5). The variable x moves along the lower line at a constant speed starting at $x = 1$. For each unit of time, we get the same change in x. The corresponding variable $y = \log x$ moves along the second line, starting at $y = 0 = \log 1$, with changes that get progressively smaller as x gets larger.

Today, instead of drawing parallel lines on which x and $\log x$ move, we place axes perpendicular to each other and simultaneously trace out the motions of x and $\log x$, producing what we know as the graph of $y = \log x$ (Figure 2.6). It is common for students to view such a graph as a static object. To understand calculus, they must see it as dynamic. Our variables x and y are changing over time, and we can view the graph as a parametric curve, $(x(t), y(t))$. The slope at any point is describing the change in y as a

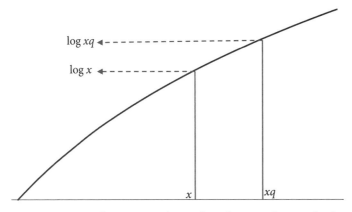

Figure 2.6. Figure 2.5 with axes at right angles, showing the standard graphical representation of $\log x$.

proportion of the change in x. It is important to remember that recognizing this constant of proportionality as a slope is a sophisticated step that would only emerge much later in the seventeenth century.

Napier investigated the ratio of the change in the logarithm to the change in x. In algebraic language, Napier used x and xq, where q is very slightly larger than 1, to denote two close values of the argument of the logarithm. Because

$$\frac{xq}{x} = \frac{q}{1},$$

equation (2.5) implies that the change in the logarithm depends only on q,

$$\log xq - \log x = \log q - \log 1 = \log q.$$

In effect, Napier was looking at the derivative of the logarithm. The ratio of the change in the logarithm to the change in x decreases as the reciprocal of x,

$$\frac{\log xq - \log x}{xq - x} = \frac{\log q}{x(q-1)} = \frac{1}{x}\frac{\log q}{q-1}.$$

Napier made the assumption that both functions start (when $x = 1$) at the same speed. The ratio of the instantaneous velocities at $x = 1$ is

$$\lim_{q \to 1} \frac{1}{1}\frac{\log q}{q-1}.$$

TABLE 2.1. Approximations to $\lim_{q \to 1} (\log_{1/r} q)/(q-1)$

r^{-1}	$\left(\log_{1/r} \left(1 + 10^{-4}\right)\right)/10^{-4}$	$\left(\log_{1/r} \left(1 + 10^{-7}\right)\right)/10^{-7}$
2	1.442620	1.442695
3	0.910194	0.910239
4	0.721311	0.721247
5	0.621304	0.621335
6	0.558083	0.558111
7	0.513873	0.513898
8	0.480874	0.480898
9	0.455097	0.455120
10	0.434273	0.434294

Napier was implicitly choosing the base for which

$$\lim_{q \to 1} \frac{\log q}{q-1} = 1.$$

He never worked out the actual value of the base that this implies. The knowledge that corresponding points on the two lines started at the same velocity was sufficient for him to work out tables of values for NapLog.[5]

For values of q that are extremely close to 1, the value of $(\log q)/(q-1)$ can be brought as close as we wish to a constant that depends only on the base of the logarithm. Table 2.1 illustrates some of the values and how they depend on the base, r^{-1}. There is a unique value of r^{-1}, lying between 2 and 3, for which the two points start at the same speed. The appropriate value of r^{-1} is 2.7182818..., the number that Euler in 1731 first designated as e. Using the fact that $\log_{1/r} x = -\log_r x$ and writing $\ln x$ for $\log_e x$, Napier's logarithm is actually

$$\text{NapLog } x = \log_{1/e} x - 7 \log_{1/e} 10 = 7 \ln 10 - \ln x.$$

Napier published his *Mirifici logarithmorum cononis constructio* (The construction of the marvelous canon of logarithms) in 1614. Henry Briggs (1561–1630), professor of geometry at Gresham College, London, was one of its first readers. He immediately saw how to improve it and headed up to Scotland where he spent a month with Napier. Briggs realized that logarithms would be more useful if they satisfied equation (2.5), implying $\log 1 = 0$, and if 10 were chosen as the base.[6] As Briggs later recalled, Napier admitted

> that for a long time he had been sensible of the same thing, and had been anxious to accomplish it, but that he had published those he had already prepared, until he could construct tables more convenient. (Havil, 2014, p. 189)

Napier died in 1617, leaving the task of constructing these more convenient tables to Briggs.

It was the Belgian philosopher Gregory of Saint-Vincent (1584–1667) and his student Alphonse Antonio de Sarasa (1617–1667) who in 1649 established that the area beneath the curve $y = 1/x$ from 1 to a had the properties of a logarithmic function,

$$\int_1^{ab} \frac{dx}{x} = \int_1^a \frac{dx}{x} + \int_1^b \frac{dx}{x}.$$

In 1668, Nicholas Mercator[7] (1620–1687) published his *Logarithmotechnia* where he coined the term *natural logarithm* for these functions. Using their representation as areas, he derived the formulas

$$\ln(1+x) = x - \frac{x^2}{2} + \frac{x^3}{3} - \frac{x^4}{4} + \cdots,$$

$$\ln(1-x) = -x - \frac{x^2}{2} - \frac{x^3}{3} - \frac{x^4}{4} - \cdots, \quad \text{and therefore}$$

$$\ln \frac{1+x}{1-x} = 2\left(x + \frac{x^3}{3} + \frac{x^5}{5} + \cdots\right),$$

facilitating the calculation of their values.

2.3
The Emergence of Algebra

Calculus is written in the language of algebra. In fact, for many students it amounts to little more than mastering manipulations of algebraic expressions. Although such facility is meaningless by itself, the reason that calculus is regarded as *the* preeminent tool for calculation is that such easily memorized procedures can be used to obtain solutions to deep and challenging problems. As we have seen, the earliest appearances of what today we recognize as differential calculus were the derivatives of the sine and natural logarithm. Derivatives of polynomials would come later, simply because they were less obviously useful. But now, as we move into the seventeenth century, we will see algebra come to play a central role.

It may seem strange to devote a section to the history of algebra within a history of calculus, but calculus relies deeply on the algebraic notation that did not fully emerge until the seventeenth century. I believe that it is worth sketching the story of how it came about. If nothing else, this clarifies the implausibility of the development of a true calculus any earlier than the seventeenth century or anywhere other than western Europe.

Algebra has ancient roots. Almost four thousand years ago, Babylonian scribes sharpened their mathematical skills on problems that today we would classify as solving quadratic equations: The length of a trench exceeds its width by 3 1/2 rods. The area is 7 1/2 sar (square rods). What are its dimensions? The method of solution is equivalent to the algebraic technique of completing the square, albeit done geometrically by constructing a square and determining how much area beyond 7 1/2 sar is needed to complete it.

Euclid included methods of solution of such problems in Book II of the *Elements*, and Diophantus of Alexandria (circa 200–284 CE) introduced the use of letters to stand for unknown quantities. But algebra really became a subject in its own right with the work of Muhammad al-Khwarizmi of Baghdad (circa 780–850). Al-Khwarizmi's algebra was restricted almost entirely to solutions of quadratic equations, but over the succeeding centuries Islamic scholars would expand the knowledge of algebra, exploring systems of linear equations and special cases of equations of higher order, as well as general methods for approximating solutions of polynomial equations.

Al-Khwarizmi, who flourished during the reign of al-Ma'mun (813–833), was one of the scholars at the House of Wisdom. He was instrumental in convincing the Islamic world to adopt the Hindu numeral system, the place value system based on the ten digits 0 through 9 that we use today. A corruption of his name, *algorism*, was used in English up to the eighteenth century to refer to this decimal system of numeration. A variation of the spelling, *algorithm*, was adopted in the twentieth century to mean a step-by-step process or procedure. The text that we today identify as the beginning of algebra, al-Khwarizmi's *The Condensed Book on the Calculation of al-Jabr and al-Muqabala*, references in its title two basic methods of maintaining a balanced equation: *al-Jabr*—from which the word *algebra* is derived—means "restoring" and refers to the transposition of a subtracted quantity on one side of an equation to an added quantity on the other; *al-Muqabala* means "comparing" and refers to subtracting equal quantities from each side of an equation.

This book contains the first fully systematic treatment of quadratic equations. Although al-Khwarizmi demonstrated the importance of working with balanced equations to solve for an unknown quantity, he described the equations and procedures entirely in words and justified his methods using geometric arguments. Because he thought of these problems geometrically, both the coefficients and solutions were necessarily positive.

The first quadratic equation that he sets out to solve is the following:

A square and 10 roots are equal to 39 units. (Al-Khwarizmi, 1915, p. 71)

By this he means: A square plus a rectangle of length 10 and width equal to the side of the square have a combined area of 39 units. In modern algebraic notation, this is the equation $x^2 + 10x = 39$. The solution is to take half the length of the rectangle, 5, and add to both sides of the equation the square of that length, 25. On one side of the equation, we are dealing with a square whose sides are five longer than the original square, $x^2 + 10x + 25 = (x+5)^2$. On the other side, we have $39 + 25 = 64$ units, the size of a square of size 8. We have transformed our equality to $(x+5)^2 = 8^2$. Therefore, the original square must have sides of length $8 - 5 = 3$.

The derivation given by al-Khwarizmi is shown in Figure 2.7. He took the rectangle and sliced it into four thin rectangles of length equal to the side of the square and width equal to $\frac{10}{4} = 2\frac{1}{2}$. He added four squares to the

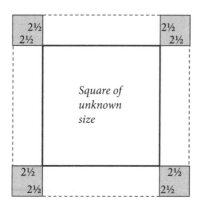

Figure 2.7. Al-Khwarizmi's geometric method for finding the root of $x^2 + 10x = 39$. After adding the four grey squares in the corners, the total area is $39 + 25 = 64$.

corners, each of area $2\frac{1}{2} \times 2\frac{1}{2} = 6\frac{1}{4}$, so that four of them equal a square of combined area 25. Adding these 25 units to the original 39 units yields a square of total area 64 whose sides are 5 longer than those of the original square.

Today's algebraic notation enables us to write this procedure as

$$x^2 + 10x = 39,$$

$$x^2 + 10x + 25 = 39 + 25 = 64,$$

$$(x + 5)^2 = 8^2,$$

$$x + 5 = 8,$$

(2.6) $$x = 8 - 5 = 3.$$

The notation improves efficiency, but something is lost if no connection is ever made to the original geometric image.

Today we would identify a second solution: $-8 - 5 = -13$. Geometrically, this makes no sense, which is why al-Khwarizmi ignored this solution. This reluctance to embrace negative solutions continued well into the seventeenth century. Descartes referred to 3 as a "true solution" of the equation $x^2 + 10x = 39$, while -13 was a "false solution." Even Newton and Leibniz shied away from negative numbers unless forced to use them.

Robert of Chester produced the first Latin translation of al-Khwarizmi's al-Jabr and al-Muqabala in 1145. Robert was an Englishman who served as archdeacon of Pamplona in Spain. He is remembered for his translations of Arabic texts into English and Latin, including the first English translation of the *Quran*. In 1202, Leonardo of Pisa—known to later historians as Fibonacci, son of the Bonacci family—published the seminal book *Liber abaci*, the book of calculations. Leonardo, a merchant of Pisa, traveled extensively through North Africa where he studied mathematics. This, his first book, focused on introducing Hindu-Arabic numerals and explained the mathematics that was useful for merchants, but it included a chapter on "Aljebra et almuchabala" that drew for its examples on the works of al-Khwarizmi and other Islamic mathematicians.[8]

Following the practice in Arabic of referring to the unknown quantity as *sha'i*, the "thing," Latin authors referred to it as *res* or *rebus*, and in Italian it became *cosa*. By the time Robert Recorde (1510–1558) published the first algebra book in English, *The Whetstone of Witte*, in 1557, algebraists were known across Europe as *cossists*. Recorde described algebra as "the arte of cosslike numbers," and his title is a play on this word since *cos* is Latin for whetstone.[9]

Algebra came into its own in 1545 with the publication of Girolamo Cardano's (1501–1576) *Artis Magnae Sive de Regulis Algebraicis* (*The Great Art or The Rules of Algebra*). For the first time, algebra moved beyond linear and quadratic equations and a few very special equations of higher order. Cardano, building on work of Scipione del Ferro, Niccolò Tartaglia, and Lodovico Ferrari, undertook a systematic treatment of the solutions of quadratic and cubic equations and included Ferrari's exact solution for equations of degree four.

In his first chapter, Cardano explained the importance of considering negative solutions, though he labeled the positive solutions as "true" and the negative solutions as "fictitious" and did not consider the possibility that 0 could be a solution.[10] The greatest inconvenience was the lack of a representation for an unspecified constant or coefficient, the *a*, *b*, and *N* in an equation such as

$$(2.7) \qquad x^3 + ax^2 + bx = N.$$

Cardano was forced to consider separately those equations where the constant was on the left or the right; the linear term was on the left, right, or did not exist; and the quadratic term was on the left, right, or did not exist. Ignoring the trivial equation $x^3 = N$ and equations that possessed no

Figure 2.8. François Viète.

positive solutions such as $x^3 + x^2 + 1 = 0$, Cardano was left with thirteen separate cases, each of which he illustrated using geometric arguments in the style of al-Khwarizmi.

The entire book was written with almost no use of symbol or abbreviation. Thus, Cardano wrote

1 *cubum p* : 8 *rebus, aequalem* 64,

meaning: A cube plus eight of the unknown quantities equals 64. Today we would represent this by

$$x^3 + 8x = 64.$$

In his foreword to the 1968 translation of Cardano's *Ars Magna*, Oystein Ore wrote,

> In dealing with problems as complicated as the solution of higher degree equations, it is evident that Cardano is straining to the utmost the capabilities of the algebraic system available to him. (Cardano, 1968, p. viii)

That changed. His successors began to introduce abbreviations and to invent symbols. One of the most influential of these algebraists was François Viète (1540–1603).

Viète was a lawyer with ties to the royal court in Paris. A Protestant during a time of religious wars in France, he served as privy counsellor to both the Catholic Henry III and the Protestant Henry IV. Most of his mathematics was done in the relatively quiet periods when court intrigue had forced him out of the inner circle. In 1591, he began the publication of *Isagoge in Artem Analyticem* (*Introduction to the Analytic Art*). One of his most influential innovations was the use of letters to represent both unknown quantities and unspecified constants or coefficients.

Today we write $y = ax$ and interpret x and y as the variables and a as the constant coefficient without even realizing that we are relying on a convention that, in fact, dates back only to René Descartes. It was Descartes who established the practice of using the letters near the end of the alphabet for the variables and those near the start of the alphabet for constants. This idea, to use letters in both senses, came from Viète, although Viète proposed to use vowels for variables and consonants as constants. Thus, that same equation might be written as $E = B A$, understanding that E and A are variables and B is a constant. While the alliteration of "vowel" with "variable" and "consonant" with "constant" (which also works in French) might be a useful mnemonic, Descartes' convention is more readily recognized.

The truly important innovation was not the use of letters for the unknowns; Diophantus had done that. It was the use of letters for the constants. Previous algebraists, in explaining a method for finding a solution to an equation, needed to choose a particular set of numerical values to illustrate the method. Viète could explain his methods in complete generality, potentially reducing Cardano's thirteen cases to the single cubic equation (2.7). Equipped with these indeterminant constants, Viète was able to take a second important step. He replaced Cardano's geometric arguments with sets of rules for maintaining balanced equations, enabling one to solve for the unknown quantity in much the way we do today.

But there were still serious drawbacks to Viète's algebra. Unlike Cardano, he did not admit negative solutions. Also, he was concerned that dimensions matched. Thus, in an equation such as

$$x^3 + ax = b,$$

x^3 is the volume of a three-dimensional cube. To maintain dimensional coherence, the coefficient a must be two-dimensional and the constant b must be three-dimensional. At the least, this introduces a concern that is often unnecessary. It also can be restrictive because x^2 must be the area of a square and cannot represent a length. Viète would never have written the equation $x^3 = x^2$. It would have to be $x^3 = 1 \cdot x^2$, the constant 1 introduced to carry the extra dimension. This continued to be a requirement throughout the seventeenth century.

Viète's notation was also archaic. Like Cardano, he did not use exponents. Thus, A^4 would be written out as *A quadrato-quadratum*, or perhaps shortened to *A quad-quad* or *A qq*, better, but still cumbersome. Contemporaries of Viète's were improving on this notation. Rafael Bombelli (1526–1572) used

$$1 \overset{6}{\smile} \text{ p. } 8 \overset{3}{\smile} \text{ Eguale à 20}$$

for what we would write as

$$x^6 + 8x^3 = 20.$$

In a similar vein, our Dutch engineer Simon Stevin wrote this expression[11] as

$$1^{\textcircled{6}} + 8^{\textcircled{3}} \text{ egales à 20.}$$

It would take another generation, until René Descartes, before algebra would take on a recognizably modern guise. But Descartes' real contribution to our story is the creation of the Cartesian plane, marrying algebra with geometry.

2.4
Cartesian Geometry

One of the greatest mathematical achievements of the seventeenth century was the appearance of what today we call analytic geometry. It embodies the connection between algebra and geometry that arises when a geometric curve is interpreted as an algebraic equation. We saw hints of this in the work of Nicole Oresme, but it emerged full blown in 1637, independently

Figure 2.9. René Descartes.

discovered and disseminated in the same year by both René Descartes and Pierre de Fermat.

They were both inspired by *The Collection* of Pappus. In the 1620s, Descartes and Fermat began to tackle some of the unproven theorems of Pappus. The theorems were not simple. Though at the time they were working on different problems, both hit upon the same route to a solution, which was to translate the geometric problem into one that could be stated in the new language of algebra. As algebraic problems, they could solve them. To accomplish the translation, they drew a horizontal axis and considered points that were a given vertical distance away, representing these distances as algebraic unknowns.

Descartes' problem came to his attention in 1631 or 1632. We know from a letter that he wrote to Isaac Beeckman in 1628 that he had for some years been thinking about the relationship between geometry and algebra. In particular, he had found a connection between solutions to a quadratic equation and the parabola. He applied his insights to the problem of Pappus with success and revealed his results in *La Géométrie*.

Descartes' Geometry was published in 1637 as one of three appendices to his *Discourse de la Methode*, or, to translate its full title, Discourse on the method of rightly conducting one's reason and of seeking truth in the sciences. The philosophical underpinnings of Aristotelian science were crumbling, and Descartes, not prone to excessive modesty, set out to establish a new foundation for scientific inquiry. This is the work in which he proclaimed, "I think, therefore I am." After explaining his scientific method, he illustrated its application in the three appendices: on optics, on meteorology, and a final appendix on geometry.

He began *La Géométrie* by explaining the algebraic notation he would use. Aside from his choice of the symbol ∞ to represent equality and a double hyphen, - - , for subtraction, it all looks very modern. One of his most important notational innovations arose in the use of numerical exponents such as x^3 for xxx. It is worth noting that the purpose of this notation was purely one of simplifying typesetting. Because both x^2 and xx require two symbols and the latter is slightly easier to typeset, it would be well into the eighteenth century before the product of x with itself was routinely written as x^2.

For Descartes, the inspiration for analytic geometry came from a problem posed by Pappus himself. Pappus described the contents of Apollonius's *Conics*, the earliest still extant derivation of the conic sections: ellipse, parabola, and hyperbola. He then mentioned a result that appears to have been well-known at the time. We take any four lines in the plane and, for each point, consider the four distances from that point to each of the lines: d_1, d_2, d_3, d_4.[12] We now consider the locus or set of points for which $d_1 d_2 = \lambda d_3 d_4$ for some constant λ. Pappus asserted without proof that this locus is a conic section and went on to muse about the locus created by six lines for which $d_1 d_2 d_3 = \lambda d_4 d_5 d_6$. He even raised the question of what might happen with more lines.[13]

To prove that the locus of solutions to $d_1 d_2 = \lambda d_3 d_4$ is a conic section, Descartes set C as a point on the locus and focused his attention on one of the lines, AB, where A is some fixed point on this line and B is the closest point to C on this line (see Figure 2.10). He denoted distance AB by x and distance BC by y. He now observed that the distance from C to any of the other lines can be expressed in terms of distances x and y, and—in modern language—he proved that the distance from C to any given line is a linear function of x and y, an expression of the form $ax + by + c$. The locus satisfies an equation of degree two,

$$y(a_1 x + b_1 y + c_1) = \lambda(a_2 x + b_2 y + c_2)(a_3 x + b_3 y + c_3).$$

Descartes then proved that four lines generate a conic section by showing that every second degree equation corresponds to an ellipse, parabola, or

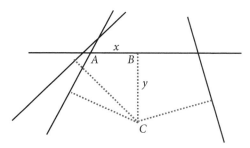

Figure 2.10. Descartes' translation of the problem of Pappus into an algebraic equation.

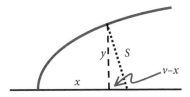

Figure 2.11. Descartes' method for finding the normal to a parabolic arc.

hyperbola (or a degenerate case such as a pair of straight lines), and every conic section can be described by an equation of degree two. He concluded with the observation that, for six lines, the points (x, y) on the locus of solutions satisfy an equation of degree three; eight lines correspond to an equation of degree four; and so on.[14]

Still less than halfway through *La Géométrie*, Descartes now tackled the problem of finding the normal to a curve (the line perpendicular to the tangent, Figure 2.11) described by an algebraic equation. Perhaps this was inspired by his interest in optics. Because the normal is perpendicular to the tangent, this is equivalent to finding the slope of the tangent, and thus marks an important watershed in the development of differential calculus.

Descartes sought the point $(v, 0)$ where the normal met the horizontal axis. He observed that if we consider a circle centered at $(v, 0)$ that is tangent to the curve at (x, y), then the line from the point of tangency to the center of the circle is the normal. The values x, y, and v satisfy the equation

$$(v - x)^2 + y^2 = s^2,$$

where s is the radius of the circle. We seek to simplify this equation, using the relationship between x and y. For example (see Figure 2.11), with the parabola $y^2 = kx$, x and v satisfy equation

$$(v - x)^2 + kx = s^2 \quad \text{or} \quad x^2 + (k - 2v)x + (v^2 - s^2) = 0.$$

This is a quadratic equation in x. If it has two distinct roots, then the circle of radius s centered at $(v, 0)$ cuts our curve twice. We will get a tangent precisely when this equation has a double solution. A quadratic polynomial with a double root at $x = r$ is of the form $x^2 - 2rx + r^2$, so this equation will have a double solution precisely when

$$x = (2v - k)/2 \quad \text{or} \quad v = x + k/2.$$

The slope of the normal at $(x, \sqrt{kx})$—the line through $(x + k/2, 0)$ and $(x, \sqrt{kx})$—is

$$\frac{\sqrt{kx}}{-k/2} = -2\sqrt{x/k}.$$

Today we would write $y = k^{1/2}x^{1/2}$ and find the slope of the tangent, the negative reciprocal of the slope of the normal,

$$\frac{dy}{dx} = \frac{1}{2}\sqrt{\frac{k}{x}}.$$

Descartes was quite enamored of this approach,

and I dare say that this is not only the most useful and most general problem in geometry that I know, but even that I have ever desired to know.[15]

There is only one problem. For most curves given by $y = f(x)$, it is nontrivial to determine the value of v for which

$$(v - x)^2 + f(x)^2 = s^2$$

has a double root. Using knowledge of calculus, if the function to the left of the equality has a double root, then its derivative at that root is zero, so we can restrict our attention to the case where

$$-2(v-x) + 2f(x)f'(x) = 0 \qquad \text{or} \qquad v = x + yf'(x).$$

But, of course, all that this establishes is that the problem of finding the value of v is equivalent to finding the slope of the tangent line.

2.5
Pierre de Fermat

In one of those coincidences that lend credence to the belief that mathematical advances have more to do with preparation of the ground than with personal insight, 1637 was also the year in which Pierre de Fermat began distributing his own account of analytic geometry.

Like Viète, Fermat was a lawyer. Born in Beaumont-de-Lomagne in the Midi-Pyrénées region of France, about 160 km southeast of Bordeaux, he spent most of his career working for the parliament in nearby Toulouse. We do not know much about his life before age 30, but by 1631 he had earned a law degree at the University of Orléans and had spent time in Bordeaux where he learned of Viète's algebra. Once settled in Toulouse, he seldom traveled, and never as far as Paris. He also never published his mathematics. His contact with other philosophers studying mathematics was entirely by correspondence, much of it via Father Marin Mersenne.

Mersenne was a Parisian monk of the Order of the Minims who was instrumental in connecting philosophers of mathematics throughout Europe, using his network of correspondents to gather and share mathematical results and arranging for these philosophers to meet each other in Paris. A close friend of Descartes, he was also responsible for publicizing much of Galileo's work.

Around 1628, Fermat picked up the Latin translation of Pappus's *The Collection* and began the task of proving the theorems that Pappus had quoted from Apollonius's *Plane Loci*. Like the problem that would start Descartes on his way toward creating Cartesian geometry, these were problems of finding all points on the plane that satisfy certain geometric conditions. As an example, Theorem II,1 places two points, A and B, on the plane and claims that the set of points, D, for which the difference of

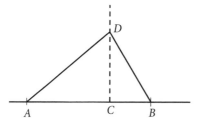

Figure 2.12. Apollonius's Theorem II,1.

the squares of the distances to A and B is equal to a given constant is a straight line perpendicular to AB (Figure 2.12).

Fermat saw that the route to solving this problem was to translate it into an algebraic equation: $AD^2 - BD^2 = \text{constant}$. We begin by finding the point C on the segment AB for which $AC^2 - BC^2 = \text{constant}$. This gives rise to a linear equation: Let c be the constant difference of the squares, d the constant distance from A to B, and x the unknown distance from A to C. Then $c = x^2 - (d - x)^2 = 2dx - d^2$. Given any point D on the perpendicular that passes through C (see Figure 2.12), the Pythagorean theorem tells us that

$$AD^2 - BD^2 = (AC^2 + CD^2) - (BC^2 + CD^2) = AC^2 - BC^2 = c.$$

Apollonius's Theorem II,5 gave him more trouble. We are given an arbitrary number of points in the plane and asked to show that a circle is the locus of points for which the sum of the squares of the distances to the given points is constant. We state this problem in the language of analytic geometry, given points $\{(a_1, b_1), \ldots, (a_n, b_n)\}$. We want to describe the locus of points (x, y) for which

$$\sum_{i=1}^{n} \left((x - a_i)^2 + (y - b_i)^2 \right) = c.$$

If we expand, divide both sides by n, and move the terms that are independent of x or y to the right, this becomes

$$x^2 - \frac{2x}{n} \sum_{i=1}^{n} a_i + y^2 - \frac{2y}{n} \sum_{i=1}^{n} a_i = \frac{c}{n} - \frac{1}{n} \sum_{i=1}^{n} a_i^2 - \frac{1}{n} \sum_{i=1}^{n} b_i^2.$$

Completing the squares on the left-hand side yields

$$\left(x - \frac{1}{n}\sum_{i=1}^{n} a_i\right)^2 + \left(y - \frac{1}{n}\sum_{i=1}^{n} b_i\right)^2$$

$$= \frac{c}{n} + \frac{1}{n^2}\left(\sum_{i=1}^{n} a_i\right)^2 + \frac{1}{n^2}\left(\sum_{i=1}^{n} b_i\right)^2 - \frac{1}{n}\sum_{i=1}^{n} a_i^2 - \frac{1}{n}\sum_{i=1}^{n} b_i^2.$$

If the right-hand side is positive,[16] this is the equation of a circle with center at

$$\left(\frac{1}{n}\sum_{i=1}^{n} a_i, \frac{1}{n}\sum_{i=1}^{n} b_i\right).$$

Fermat began his study of this problem without any knowledge of analytic geometry. It took him six or seven years, until 1635, to discover a proof. He first resolved Theorem II,5 for the case of two points. After much effort, he produced a proof for an arbitrary number of points that all lie along a single line, which we can assume to be horizontal. The center of the circle will lie on this line at the point x for which the sum of the distances to those points to the left of x is equal to the sum of the distances to those points to the right of x.[17]

Fermat now tackled the general case in which the given points do not all lie on this horizontal line. He first projected each of the points onto the horizontal line (in modern language, considering just the x-coordinates) and found the center of these points. Then he constructed the perpendicular at this center, projected each of his points onto this vertical line (now noting the y-coordinates), found the center of these points, then argued algebraically that this is the center of the circle that satisfies the requirement.

By the time he had completed his proof, Fermat realized that he had a general and very powerful method for solving problems of geometric loci: projecting points onto a horizontal axis and decomposing their location into a horizontal distance from some fixed point at the left end of the segment and the vertical distance to the line.

It is important to recognize that neither Descartes nor Fermat worked with a true coordinate geometry. Boyer describes it as "ordinate

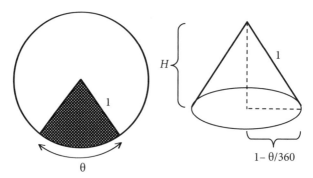

Figure 2.13. A cone of radius $1 - \theta/360$ and height $H = \sqrt{1 - (1 - \theta/360)^2}$ constructed from a circle of radius 1.

geometry,"[18] Mahoney as "uniaxial."[19] Points are described by their *abscissa* or coordinate on the horizontal axis and their vertical distance from the abscissa. Even the abscissa is described as a distance from a fixed point. Because all distances are positive, we are effectively working just in the first quadrant. This would be true of analytic geometry right through the seventeenth century, including the work of Leibniz and Newton.

It was at this time, around 1636, that Fermat began to correspond with Fr. Mersenne as well as Gilles Personne de Roberval, who had arrived in Paris in 1628 and become part of Mersenne's inner circle. Shortly thereafter, Fermat sent two manuscripts to Mersenne and Roberval: *Introduction to Plane and Solid Loci*, which explained his method for solving locus problems through translation into algebraic expressions, and his *Method for Determining Maxima and Minima and Tangents to Curved Lines*.

Back in 1629, while still in Bordeaux, Fermat had learned of a problem that Mersenne was circulating, that of finding the cone of greatest volume that could be cut from a circle of given area (see Figure 2.13). From his solutions to similar problems, we can reconstruct in modern notation how he would have solved it. If we cut a sector of angle θ (measured in degrees) from a circle of radius 1, what remains can be folded into a cone whose base has radius $1 - \theta/360$ and whose height is $\sqrt{1 - (1 - \theta/360)^2} = (1/360)\sqrt{7200\theta - \theta^2}$. The volume of the cone is

$$\frac{\pi}{3}\left(1 - \frac{\theta}{360}\right)^2 \frac{1}{360}\sqrt{7200\theta - \theta^2} = \frac{\pi}{3 \cdot 360^3}(360 - \theta)^2\sqrt{7200\theta - \theta^2}.$$

Maximizing this quantity for $0 < \theta < 360$ is equivalent to maximizing

$$(360 - \theta)^4(720\theta - \theta^2) = (360 - \theta)^4\theta(720 - \theta).$$

If we replace θ by $360 - \phi$ (a clever simplification), the quantity to be maximized is

$$\phi^4(360 - \phi)(360 + \phi) = 360^2\phi^4 - \phi^6.$$

Fermat now relied on an observation reminiscent of Descartes' method for finding the normal to a curve. If c is the maximum value of this polynomial, then the equation $360^2\phi^4 - \phi^6 = c$ must have a double root. We consider two values, ϕ and ψ, at which this expression takes on the same value,

$$360^2\phi^4 - \phi^6 = 360^2\psi^4 - \psi^6.$$

We move everything to the left side of the equation,

$$360^2(\phi^4 - \psi^4) - (\phi^6 - \psi^6) = 0,$$

and observe that for any pair of unequal roots, we can divide by $\phi - \psi$,

$$360^2(\phi^3 + \phi^2\psi + \phi\psi^2 + \psi^3) - (\phi^5 + \phi^4\psi + \phi^3\psi^2$$
$$+ \phi^2\psi^3 + \phi\psi^4 + \psi^5) = 0.$$

We find the double root by setting $\psi = \phi$ and solving the resulting equation,

$$4 \cdot 360^2\phi^3 - 6\phi^5 = 0; \quad \phi = \frac{\sqrt{6}}{3}\,360; \quad \theta = \left(1 - \frac{\sqrt{6}}{3}\right)360.$$

The maximum volume is $2\pi\sqrt{3}/27$. Viète had used exactly this process, which he called *syncrisis*, to find multiple roots of polynomials.[20] It formed the basis of Fermat's method for finding maxima and minima.

If we let $f(\phi)$ denote the polynomial in ϕ, then what he has done is to simplify the left side of

$$\frac{f(\phi) - f(\psi)}{\phi - \psi} = 0,$$

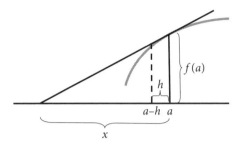

Figure 2.14. Finding the subtangent, x.

to set $\psi = \phi$, and then to solve for ϕ, a process equivalent to setting the derivative of f equal to 0 and solving for ϕ.

As he developed his method, Fermat came to refer to the second point by its distance from the first, x and $x - h$. The process involved setting up what he called an *adequality*, $f(x) \approx f(x - h)$, which could be simplified as if it were an equality, divided through by h, then solved for x by setting $h = 0$. Today we regard what he was doing as finding the limit of

$$\frac{f(x) - f(x - h)}{h}$$

as h approaches 0, accomplished by simplifying this quotient, setting $h = 0$, and then solving for x. However, for Fermat there is no indication of a limit. He gave no justification for this process, a lacuna that was very apparent to his contemporaries.

Fermat used this method to find the tangent to a curve. For Fermat, as for his contemporaries, the tangent line is determined by two points: a point on the curve and a point on the horizontal axis. The problem is to find the horizontal distance, known as the *subtangent*, to the point of intersection of the tangent and the horizontal axis. Drawing on modern functional notation to simplify the explanation, let $(a, f(a))$ be the point on the curve and x the length of the subtangent. We assume that the function is increasing at a. Using similar triangles, the vertical distance from $a - h$ to the tangent line is $f(a)(x - h)/x$. If h is close to 0, this will be close—though not equal—to $f(a - h)$ (see Figure 2.14).

He set up an adequality between $f(a)(x - h)/x$ and $f(a - h)$. From a modern perspective, Fermat sought the x for which

$$\lim_{h \to 0} \frac{f(a)(x-h)/x - f(a-h)}{h} = \frac{-f(a)}{x} + \lim_{h \to 0} \frac{f(a) - f(a-h)}{h} = 0.$$

In practice, his method for finding the subtangent was to expand $f(a)$ $(x - h) - xf(a - h)$, divide by h, then set $h = 0$ and solve for x in terms of a. This works. The limit is $-f(a)/x + f'(a)$, which is 0 when x is the subtangent, but Fermat's derivation was far from clear. His approach was viciously criticized by Descartes who described it as proceeding "*à tatons et par hasard*" (through fumbling and luck).[21]

2.6
Wallis's *Arithmetic of Infinitesimals*

By the 1650s there was a large community busily extending the techniques of calculus. The foundations had been laid by Cavalieri and Torricelli in Italy; Fermat, Mersenne, and Roberval in France; Descartes in the Netherlands; and Gregory of Saint-Vincent and de Sarasa in Belgium. New entrants included Blaise Pascal (1623–1662) who, through his father Étienne, had been introduced to the Parisian circle of mathematics. In addition to Huygens, two other students of Van Schooten, Johannes Hudde (1628–1704) and Hendrick van Heuraet (1634–1660), joined in the expansion of Descartes' insights into geometry. They were in close contact with René de Sluse (1622–1685), a Belgian also exploring questions of quadratures and tangents. Across the channel in England, John Wallis (1616–1703), William Brouncker (1620–1684), Christopher Wren (1632–1723), and William Neile (1637–1670) competed for priority in discovering formulas for areas under curves and a method for computing arc lengths.

By this time, everyone had accepted Descartes' method of pairing algebraic equations with geometric curves. Most had embraced Torricelli's version of Cavalieri's indivisibles, if not as a rigorous foundation then at least as a means of discovering new formulae. The competition was often fierce and bitter. In the midst of this activity, John Wallis emerged with his *Arithmetica Infinitorum* (Arithmetic of infinitesimals), published in 1656. As the title suggests, Wallis turned these investigations to a new direction by shifting the emphasis from geometry to arithmetic.

Figure 2.15. John Wallis.

Wallis seems to have stumbled into mathematics. As he recounted in his autobiography,[22] he knew no arithmetic until the age of 15 when his younger brother explained "Addition, Substraction [*sic*], Multiplication, Division, the Rule of Three, the Rule of Fellowship, the Rule of False-Position, [and] the Rules of Practise [*sic*] and Reduction of Coins."[23]

He continued to pick up what mathematics he could, reading Oughtred's *Clavis mathematicae* (The key to mathematics), his first introduction to algebra. At Cambridge he included among his studies Astronomy and Geography as well as "other parts of Mathematicks; though, at that time, they were scarce looked upon, with us, as Academical Studies then in fashion."[24] Ordained to the clergy in 1640, he was active on the side of the Parliamentarians in the English Civil War. In 1642, the first year of war, he decoded an intercepted message between Royalists. It was a simple substitution code that he cracked in an evening. His reputation spread, and he found himself called upon to break increasingly more difficult ciphers.

Oliver Cromwell had dismissed the Savilian Chair of Geometry at Oxford, Peter Turner, a Royalist. In 1649 he appointed Wallis in his place. In Wallis's own words, "Mathematicks which had before been a pleasing Diversion, was now to be my serious Study."[25]

Wallis had been out of the loop. He found and read Descartes' *Géometrie* and Torricelli's *Opera geometrica*. But knowing nothing of the unpublished work of Fermat or Roberval, he had to rediscover much of what was already known. He produced *De sectionibus conicis* (On conic sections) in 1652, and in 1656 published his results on the general problem of areas and volumes in *Arithmetica Infinitorum*. This is where the symbol he invented for infinity, the "lazy eight," ∞, makes its first appearance.

Wallis was heavily influenced by Torricelli, as was clear in his very first result, the determination of the area of a triangle as one half the base times the height. He first established the arithmetic rule that

$$\frac{0+1+2+\cdots+l}{l+l+l+\cdots+l} = \frac{(l+1)l/2}{(l+1)l} = \frac{1}{2}.$$

He then claimed that a triangle consists "of an infinite number of parallel lines in arithmetic proportion"[26] (see Figure 1.16 in section 1.6), while a parallelogram of the same base and height consists of an equal number of lines all of length equal to the base. The ratio of the areas therefore is also one half.

The methods of Cavalieri and Torricelli were still regarded with deep suspicion, and Wallis's contributions came in for a great deal of criticism, compounded by the fact that sometimes Wallis described the decomposition as into lines, other times as into infinitesimally thin parallelograms. Thomas Hobbes (1588–1671) was particularly scathing. Either the lines have no breadth, and so the height consists "of an infinite number of nothings, and consequently the area of your triangle has no quantity,"[27] or these are actual parallelograms, in which case the figure is not a triangle.

Wallis himself was aware that great care needed to be taken in regarding the triangle as an infinite union of infinitesimally thin parallelograms, pointing out in the comment following Proposition 13 that although this representation could be used to determine the area of the triangle, it could lead to error if one attempted to use it compute the perimeter.[28] Nevertheless, he plunged ahead, showing that for any positive integer k,

$$\frac{0^k+1^k+2^k+\cdots+l^k}{l^k+l^k+l^k+\cdots+l^k} = \frac{(l+1)l^k/k}{(l+1)l^k} + \text{remainder} = \frac{1}{k} + \text{remainder},$$

Figure 2.16. James Gregory.

where the remainder approaches 0 as l increases (Proposition 44).[29] He then extended this result to the case where k is an arbitrary nonzero rational exponent, positive or negative. In the process, he became the first person to use rational and negative exponents. He used this insight to find the area under an arbitrary curve of the form $y = cx^k$ and showed how this could be applied to solve a wide variety of problems of areas and volumes.

As we have seen, others had discovered these integration formulae. Wallis's most original contribution in this book was his derivation of the infinite product formula for π,[30]

$$(2.8) \qquad \pi = 4 \cdot \frac{2}{3} \cdot \frac{4}{3} \cdot \frac{4}{5} \cdot \frac{6}{5} \cdot \frac{6}{7} \cdot \frac{8}{7} \cdots .$$

Isaac Newton read Wallis's *Arithmetica Infinitorum* in the winter of 1664–65, his last year as a student at Cambridge. It profoundly influenced his development of calculus, and, as Newton later explained in a letter to Leibniz (October 24, 1676), Wallis's derivation of the product formula for π led directly to his own discovery of the general binomial theorem,

$$(1 + x)^k = 1 + \frac{k}{1}x + \frac{k(k-1)}{2!}x^2 + \frac{k(k-1)(k-2)}{3!}x^3 + \cdots$$

for any rational value of k.

The Scottish philosopher James Gregory (1638–1675) traveled to Padua in 1664 to study with Stefano degli Angeli (1623–1697), another student of Cavalieri. In 1668 when Gregory left Italy to take up the new chair of mathematics at the University of Saint Andrews in Scotland, he published his great work on calculus, *Geometriæ pars universalis* (*The Universal Part of Geometry*). Among the many accomplishments in this book is the first complete statement of the Fundamental Theorem of Integral Calculus.[31] As explained in the next paragraph, it is buried within his study of the calculation of arc lengths, and there is no indication that he recognized its full significance.

In 1657, William Neile at Oxford and Van Heuraet in the Netherlands independently discovered the formula for the arc length of the semi-cubical parabola, $y = x^{3/2}$, setting up a question of priority that would be disputed by Wallis and Huygens, their respective champions, for many years.[32] Although their work suggested how to approach this problem of "rectification of curves" in general, Gregory was the first to work it out in full generality. The result is that the length of the curve $y = f(x)$ from $x = a$ to $x = b$ is equal to the area under the curve $y = \sqrt{1 + (f'(x))^2}$ over the same interval. Gregory then posed the following question: If we know that the length of the curve $y = f(x)$ is given by the area under the graph of $y = g(x)$ and we know the function g, can we find f? In modern notation, this amounts to solving the differential equation

$$g(x) = \sqrt{1 + (f'(x))^2}, \quad \text{or} \quad f'(x) = \sqrt{(g(x))^2 - 1}.$$

Because we know that it is safe to manipulate these as if they were ordinary algebraic equations, the equivalence appears simple. But to Gregory, they were very different geometric statements. To establish their equivalence, he needed the fact that the rate of change of an accumulator function is the ordinate of the function that is being accumulated. In other words, he needed to establish geometrically that

$$\frac{d}{dx} \int_a^x f(t)\, dt = f(x).$$

Another significant accomplishment of Gregory was to share with Newton the discovery of Taylor series. In 1671 Gregory wrote to John

Collins explaining the derivation of these series, giving examples that included the Taylor series expansions of tan x, arctan x, sec x, and ln (sec x). He learned that two years earlier, in 1669, Isaac Newton had circulated the manuscript *On analysis by equations with an infinite number of terms*, which included a description of these series. Thinking he had been scooped, Gregory refrained from publishing his results. Four years later, at the age of 36 and the height of his powers, Gregory died suddenly of a stroke. Had he lived, his greatness might have rivaled that of Newton and Leibniz.

Unfortunately, Newton never published his work on infinite series. The entire subject became part of the knowledge of those working in calculus, shared in letters and through personal contact, but not appearing in print until 1715 when Brook Taylor (1685–1731) explained it in *Methodus incrementorum directa et inversa* (Direct and indirect methods of incrementation). In section 3.2 I will describe how Gregory and Newton came to discover Taylor series.

I conclude this section with a brief nod to Isaac Barrow (1630–1677), the Lucasian Professor of Mathematics at Cambridge while Newton was a student there (1660–1664) and the author of *Lessons in Geometry*, published in 1669 with assistance from Newton. *Lessons in Geometry* is a culmination of the work of the previous half century on tangents, areas, and volumes.

Interpretations of Barrow's influence on Newton have waxed and waned. In the early twentieth century, J. M. Child saw him as the true originator of calculus. Today, his influence is considered minimal. *Lessons in Geometry* does contain a statement of the Fundamental Theorem of Integral Calculus, but Barrow presented it as an aside of no great importance. And although Barrow may have been Cambridge's only professor of mathematics at the time Newton was a student, he was never Newton's teacher or tutor. Newton learned his mathematics by reading the classics of the time, especially Oughtred on algebra, Descartes on geometry, and Wallis on infinitesimal analysis. Barrow enlisted Newton's help in getting his book out the door because he was done with mathematics. That year he resigned his chair at Cambridge to take up the position of royal chaplain to Charles II.

Lessons in Geometry looked backward. By 1669, Newton was looking forward.

2.7
Newton and the Fundamental Theorem

Although accumulation problems were well understood and most of the techniques of integral and differential calculus had been discovered long before Newton or Leibniz entered the scene, it was far from a coherent theory. The genius of Newton and Leibniz, and the reason that they are credited as the founders of calculus, is that they were the first to understand and appreciate the full power of understanding integration and differentiation as inverse processes.[33] This is the crux of the Fundamental Theorem of Integral Calculus, usually referred to today as simply the Fundamental Theorem of Calculus.[34]

During Newton's two years back in Lincolnshire following his graduation from Cambridge, he organized his thoughts on calculus. In October 1666 he summarized his discoveries in a manuscript known as the *Tract on Fluxions* that was intended to be shared with John Collins.[35] To Newton, *fluxion* was a rate of change over time. Throughout his mathematical work, quantities that varied did so as functions of time. This is important. When Newton sought to find the slope of the tangent line to the curve $y = x^3$ at a point, he *always* treated this as implicit differentiation, $\frac{dy}{dt} = 3x^2 \frac{dx}{dt}$. The slope of the tangent line is the ratio of the rates of change of the y and x variables,

$$\frac{dy}{dx} = \frac{dy/dt}{dx/dt} = \frac{3x^2 \frac{dx}{dt}}{\frac{dx}{dt}} = 3x^2.$$

This may seem cumbersome, but it conveys an important point that is often lost on calculus students: The derivative is telling us much more than the slope of a tangent line. It encodes the relationship of the rates of change of the two variables.

Two problems stand out from the 1666 tract, the fifth and seventh:

Prob 5^t. To find the nature of the crooked line [curve] whose area is expressed by any given equation. That is, the nature of the area being given to find the nature of the crooked line whose area it is. (Newton, 1666, p. 427)

Prob: 7. The nature of any crooked line being given to find its area when it may bee [*sic*]. (Newton, 1666, p. 430)

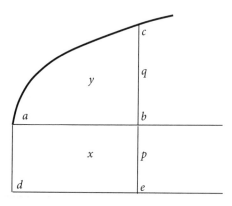

Figure 2.17. Newton's illustration for his fifth and seventh problems.

These problems are accompanied by Figure 2.17. To orient the reader, the line ab is the horizontal axis above which we see the graph of the function, given by the curve ac. In the fifth problem, Newton assumed that we have a known accumulator function that describes the area under ac as a function of the abscissa, b. He denoted this function by y. The problem was to find the functional equation that describes the curve. Newton observed that the ordinate, q, is equal to the rate at which the area is increasing. In modern notation, this is the observation that

$$q = f(b) = \frac{dy}{db} = \frac{d}{db} \int_a^b f(x)\, dx,$$

the antiderivative part of the Fundamental Theorem of Integral Calculus.

The seventh problem leads to the evaluation part of the Fundamental Theorem. Now we are given the functional equation describing the curve, $q = f(b)$, and need to find the accumulator function, that function of b that describes the area under the curve up to the point b. Newton referred back to his solution to the fifth problem. If he could find a function $y(b)$ whose derivative is $f(b)$, then the area is given by $y(b)$, which is equal to $y(b) - y(a)$ since he has assumed that $y(a) = 0$. With great enthusiasm, he now presented two curves beneath which he could determine the area

$$y = \frac{ax}{\sqrt{a^2 - x^2}} \quad \text{and} \quad y = \sqrt{\frac{x^3}{a} - \frac{e^2 b}{x\sqrt{ax - x^2}}}.$$

I need to explain the part of Newton's illustration under the horizontal axis, the x, d, e, and p, because they provide a clue to some of the conceptual issues with which Newton and his contemporaries were wrestling and which can present obstacles to our own students. The difficulty involves work with ratios.

To Hellenic and Hellenistic philosophers, and by extension to the philosophers of the seventeenth century, ratios could only be ratios of like quantities: lengths to lengths, areas to areas, volumes to volumes. That fact is important in understanding how they thought about velocity. Velocity was never seen as a ratio because distance and time are incommensurable. In modern notation, the solution of Newton's fifth problem involves finding the rate at which y is changing with respect to b, the limit of $\Delta y/\Delta b$. But Δy is an area and Δb is a length. That cannot be done. The solution is to trace a second curve, actually the straight line de, located constant distance p from the horizontal axis, and let $x = x(b)$ denote the accumulated area.

The idea of a negative coordinate never occurred to Newton or Leibniz. Both points c and e have positive ordinates. Today we would draw the line de above the horizontal axis, but that would greatly complicate the drawing.

Now we can take a ratio of like quantities,

$$\frac{\Delta y}{\Delta x} \approx \frac{q \cdot \Delta b}{p \cdot \Delta b} = \frac{q}{p},$$

with the approximation approaching equality as Δb approaches 0. The quantity p is an arbitrary constant. Today we would be inclined to set $p = 1$, which means that $x = b$, and therefore $dy/db = q$. We can get away with this if we are careful about our units. The rate of change of y is measured in area per unit time. The rate of change of b is measured in length per unit time. Area/time $\div$ length/time = length. The scientists of the eighteenth century realized that as long as they were careful with units, they could dispense with restricting ratios to like quantities. But pedagogical problems can arise when students start using expressions such as $dy/db = q$ with no regard to the unit that indicates what is being measured. Newton and Leibniz were overly cautious. We may have moved too far the other way.

2.8
Leibniz and the Bernoullis

Gottfried Leibniz (1646–1716) was trained as a lawyer and worked as private secretary to Baron von Boyneburg, a position that often required travel on diplomatic missions. It was in Paris in 1672 that he met Christiaan Huygens, probably the foremost Continental expert on calculus at that time. When he arrived in Paris, Leibniz had an inflated sense of his mathematical abilities, which only increased when he succeeded in solving a problem that Huygens posed to him. This was to sum the reciprocals of the triangular numbers,

$$(2.9) \qquad 1 + \frac{1}{3} + \frac{1}{6} + \frac{1}{10} + \frac{1}{15} + \frac{1}{21} + \cdots = \sum_{n=1}^{\infty} \frac{2}{n(n+1)}.$$

Leibniz realized that he could use partial fraction decomposition to write each fraction as a difference,

$$\frac{2}{n(n+1)} = \frac{2}{n} - \frac{2}{n+1}.$$

By cancellation, the sum equals

$$\frac{2}{1} - \frac{2}{2} + \frac{2}{2} - \frac{2}{3} + \frac{2}{3} - \frac{2}{4} + \cdots = 2.$$

The following year Leibniz traveled to England. He tried to impress the English philosophers with this result, only to discover that it had been published by Pietro Mengoli more than two decades earlier. Leibniz also began to realize how little he knew about recent developments in calculus and came away with a copy of Barrow's *Lessons in Geometry*.

For the next several years, Leibniz lived in Paris and began the serious study of mathematics under the tutelage of Huygens. By the fall of 1673, he had rediscovered the Fundamental Theorem of Integral Calculus. Over the years 1673–1676 he worked out for himself the techniques for applying the rules of calculus, including a sophisticated understanding of the role of integration by substitution and by parts. He relied on the language of infinitesimals and created an appropriate notation that still serves us

Figure 2.18. Gottfried Leibniz.

well today. He invented dy/dx for the derivative, understood as a ratio of infinitesimals, and $\int y\,dx$ for the sum of products of the form $y\,dx$.

Leibniz referred to his differentials as infinitesimals, but he was clear that they were a mathematical fiction, a shorthand for describing quantities that could be made as small as one wished. In a letter to Bernard Nieuwentijdt, he explained his thinking:

> When we speak of infinitely great (or more strictly unlimited), or of infinitely small quantities (i.e., the very least of those within our knowledge), it is understood that we mean quantities that are indefinitely great or indefinitely small, i.e., as great as you please, or as small as you please, *so that the error that any one may assign may be less than a certain assigned quantity.*. . . If any one wishes to understand these [the infinitely great and infinitely small] as the ultimate things, or as truly infinite, it can be done, and that too without falling back upon a controversy about the reality of extensions, or of infinite continuums in general, or of the infinitely small, ay, even though he think that such things are utterly impossible; it will be sufficient simply to make use of them as a tool that has advantages for the purpose of the calculation just as the algebraists retain imaginary roots with great profit. (Child, 1920, p. 150; italics added)

The italicized portion of the quotation points to Leibniz's insistence that infinitesimals are merely suggestions of differences that can be brought arbitrarily close to zero.

In some sense, Leibniz's greatest accomplishment was to get this knowledge to where it could be appreciated by others. In 1682, he helped to found *Acta Eruditorum*, the first scientific journal from what is today Germany and one of the very first scientific journals in Europe.[36] Leibniz published the first account of his work on calculus in this journal in 1684.[37] It attracted the attention of two Swiss brothers with a penchant for mathematics, Jacob and Johann Bernoulli (1654–1705 and 1667–1748, respectively).

The Bernoulli brothers were much more inclined to embrace infinitesimals as actual quantities, to accept Leibniz's assurance that it was safe to treat them as such. They did so with great success. Over the period 1690 to 1697, they demonstrated their mastery of the techniques of calculus by using differentials to find curves with special properties. These included:

- The curve of uniform descent: The curve along which a rolling ball descends with a constant vertical velocity as it accelerates under the pull of gravity.
- The isochrone curve: The curve along which the time it takes for the ball to reach the bottom is independent of the position on the curve where the ball is placed.
- The brachistochrone curve: The curve connecting points A and B (with B lower than A) which minimizes the time it takes a ball to roll from A to B.
- The catenary curve: The curve that describes how a heavy rope or chain hangs.

For each of these problems, one has information about the slope of the curve at each point. The Bernoullis used this information to construct a differential equation that described the solution.

The older Bernoulli brother, Jacob, had secured the only professorship of mathematics at the University of Basel, their home town. The younger brother, Johann, was forced to seek employment elsewhere. This was not a simple matter.

Figure 2.19. Leonhard Euler.

In 1691, Johann traveled to Paris where he met the Marquis Guillaume François Antoine de l'Hospital (1661–1701). The marquis was a competent mathematician and eager to learn the new calculus. In 1694, he sent Bernoulli an interesting proposal, a retainer of 300 pounds (thirty times the annual salary of an unskilled laborer of the time) in exchange for allowing l'Hospital to publish Bernoulli's mathematical discoveries under his own name. The resulting book, *Analyze des infiniments petits*, became the first to explain Leibniz's calculus. It included the technique for finding limits of quotients that approach 0/0 that would come to be known as l'Hospital's rule.[38]

It is not clear when the ∞/∞ version of l'Hospital's rule was discovered. It can be found in Cauchy's lessons in analysis from the 1820s. That may have been the first time it was published.

2.9
Functions and Differential Equations

The earliest use of the term "function" in anything approaching its modern mathematical sense is found in the correspondence between Leibniz and Johann Bernoulli in the 1690s, though it meant nothing more than an unspecified quantity that had been calculated, rather than the rule by which it was computed. By 1718, Bernoulli defined this term to mean the rule for calculating such a quantity. The term was embraced by Bernoulli's student, Leonhard Euler (1707–1783). Its definition occurs near the beginning of his 1748 *Introduction to Analysis of the Infinite.*

> A function of a variable quantity is an analytical expression composed in any way from this variable quantity and from numbers or constant quantities. (Euler, 1988, p. 3)

Seven years later, in *Foundations of Differential Calculus,* he clarified both the breadth of his definition and the intent, that the purpose of a function is to connect two varying quantities.

> Those quantities that depend on others in this way, namely, those that undergo a change when others change, are called *functions* of these quantities. This definition applies rather widely and includes all ways in which one quantity can be determined by others. Hence, if x designates the variable quantity, all other quantities that in any way depend on x or are determined by it are called its functions. (Euler, 2000, p. vi)

In this simple definition, Euler marked an important shift in our understanding of differential calculus from a geometric to a functional interpretation. Functional relationships were freed from the need to be encoded geometrically, giving far greater rein to the kinds of situations that could be modeled and opening up all of the possibilities inherent in differential equations.

Leonhard Euler grew up in Basel, Switzerland, and attended its university with the intention of preparing for the ministry. Jacob Bernoulli died in 1705, two years before Euler's birth. Jacob's younger brother, Johann, took over the chair in mathematics. When Euler entered the University of Basel, Johann Bernoulli quickly recognized and encouraged his mathematical talent. In 1727, Euler followed Johann's son Daniel (1700–1782), another very talented mathematician and a close friend, in moving to Saint Petersburg, Russia, to take up a position in the newly created Saint Petersburg Academy. He left Russia for Berlin in 1741, where King Frederick II, known as Frederick the Great, appointed him director of his Astronomical Observatory.

Frederick collected some of the most brilliant philosophers of the era, including Voltaire, Maupertuis, Diderot, and Montesquieu, with whom he enjoyed engaging in brilliant conversation. He was deeply disappointed with Euler who lacked their flair. Euler, in turn, was frustrated by the time-consuming tasks to which he was assigned. In 1766, Euler was brought back to Saint Petersburg by Catherine II, Catherine the Great, where he was able to fully engage in the projects that most interested him. In his early 30s Euler had developed an infection that led to blindness in one eye. His other eye gradually weakened, and for the last twelve years of his life, he was totally blind, but this did little to slow down his prodigious output.

No one epitomizes the golden age of eighteenth-century calculus better than Euler. No one has been more prolific. His collected works run to eighty quarto volumes of 300–600 pages each, with an additional four volumes that have yet to be published. His interests were broad, and he valued clear exposition as highly as original discovery.

He set the standard for published mathematics, writing in a style that looks modern to our eyes, choosing his notation with care and laying out the argument in a pattern of text interspersed with displayed equations. It was Euler who popularized the use of π to denote the ratio of the circumference of a circle to its diameter. He also was the first to employ the letter e to denote the base of the natural logarithm. While exponential notation had been around since Descartes, it was Euler who introduced the exponential function, a^x, the inverse of the logarithmic function for the base a. He then set the precedent of defining the logarithm as the inverse of the exponential function.

Euler revealed the power of calculus to the scientific community. In his systematic and successful applications to problems in mechanics, hydrodynamics, and astronomy he not only solved important problems, he laid the foundations for future generations of scientists to build the mathematical infrastructure that is still employed today.

Euler was a master of differential equations, both in recognizing how they could be used to model physical situations and in finding solutions to these equations. Nowhere is this clearer than in his work on fluid mechanics. In *Principia motus fluidorum* (Principles of the motion of fluids), written in 1752,[39] he built from two-dimensional flow to three-dimensional flow, then added time as a variable, relying on a function from four variables $(x, y, z, t$; three variables describing position and one of time) to three variables $(u, v, w$; the three components of the fluid's velocity at that position and time). In the case of a constant flow of what would come to be called an *incompressible fluid* (the flow into each region must equal the flow out), he proved, when translated into the modern notation of partial derivatives,[40] that

$$\frac{\partial u}{\partial x} + \frac{\partial v}{\partial y} + \frac{\partial w}{\partial z} = 0.$$

Today we know this as the divergence theorem. It is foundational to much of modern physics, not just in fluid and aerodynamics but also to heat transfer and the equations of electricity and magnetism.

Although his argument is a little complicated, it is worth presenting in the special case of two-dimensional flow because it gives insight into how the language of functions facilitated his analysis. For ease of comprehension, I will use the modern notation of partial derivatives. Euler couched his analysis in terms of differentials, which he thought of as infinitesimals. Although he recognized the problems inherent in treating them as zeros with well-defined ratios, he also saw the power they provided.

He began by considering the movement of a small triangular piece of the fluid with vertices at (x, y), $(x + dx, y)$, and $(x, y + dy)$ as it flows with a velocity vector that depends on the position, $\langle u(x, y), v(x, y) \rangle$. As we change the position at which we measure the velocity, the velocity components change by (see Figure 2.20)

$$du = \frac{\partial u}{\partial x} dx + \frac{\partial u}{\partial y} dy,$$

$$dv = \frac{\partial v}{\partial x} dx + \frac{\partial v}{\partial y} dy.$$

If the flow at (x, y) is given by $\langle u, v \rangle$, then we can describe the flow at $(x + dx, y)$ as

$$\left\langle u + \frac{\partial u}{\partial x} dx, v + \frac{\partial v}{\partial x} dx \right\rangle,$$

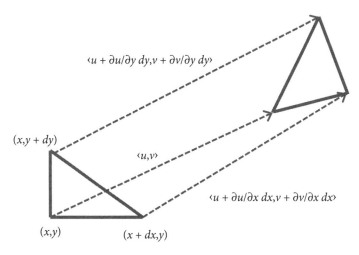

Figure 2.20. The two-dimensional movement of a small triangular piece of fluid. Solid triangles represent the initial and final position. Dashed arrows show flow vectors.

and the flow at $(x, y + dy)$ as

$$\left\langle u + \frac{\partial u}{\partial y} dy, v + \frac{\partial v}{\partial y} dy \right\rangle.$$

A short time later, dt, the points marking the vertices of our triangle have moved as follows:

$$(x, y) \longrightarrow (x + u\,dt, y + v\,dt),$$

$$(x + dx, y) \longrightarrow \left(x + dx + \left(u + \frac{\partial u}{\partial x} dx \right) dt, y + \left(v + \frac{\partial v}{\partial x} dx \right) dt \right),$$

$$(x, y + dy) \longrightarrow \left(x + \left(u + \frac{\partial u}{\partial y} dy \right) dt, y + dy + \left(v + \frac{\partial v}{\partial y} dy \right) dt \right).$$

Euler now argued that, because the lengths are infinitesimal, the image of the original triangular region of the fluid can be taken as the triangular region determined by the three new vertices.

The area of the original triangle is $\frac{1}{2} dx\, dy$. The triangle to which it is carried is spanned by the vectors

$$\left(dx + \frac{\partial u}{\partial x} dx\, dt,\; \frac{\partial v}{\partial x} dx\, dt \right), \quad \text{and} \quad \left(\frac{\partial u}{\partial y} dy\, dt,\; dy + \frac{\partial v}{\partial y} dy\, dt \right).$$

This triangle has area

$$\frac{1}{2} \left[\left(dx + \frac{\partial u}{\partial x} dx\, dt \right) \left(dy + \frac{\partial v}{\partial y} dy\, dt \right) - \left(\frac{\partial v}{\partial x} dx\, dt \right) \left(\frac{\partial u}{\partial y} dy\, dt \right) \right]$$

$$= \frac{1}{2} \left[dx\, dy + \frac{\partial u}{\partial x} dx\, dy\, dt + \frac{\partial v}{\partial y} dx\, dy\, dt + \frac{\partial u}{\partial x} \frac{\partial v}{\partial y} dx\, dy\, dt^2 - \frac{\partial v}{\partial x} \frac{\partial u}{\partial y} dx\, dy\, dt^2 \right]$$

$$= \frac{1}{2} dx\, dy + \frac{1}{2} \left(\frac{\partial u}{\partial x} + \frac{\partial v}{\partial y} + \frac{\partial u}{\partial x} \frac{\partial v}{\partial y} dt - \frac{\partial v}{\partial x} \frac{\partial u}{\partial y} dt \right) dx\, dy\, dt.$$

Because the fluid is incompressible, the initial and final areas must be the same, and therefore

$$\frac{\partial u}{\partial x} + \frac{\partial v}{\partial y} + \frac{\partial u}{\partial x} \frac{\partial v}{\partial y} dt - \frac{\partial v}{\partial x} \frac{\partial u}{\partial y} dt = 0.$$

Euler now treated dt as zero, leaving us with the divergence equation for a two-dimensional flow,

$$(2.10) \qquad\qquad \frac{\partial u}{\partial x} + \frac{\partial v}{\partial y} = 0.$$

He then considered the three-dimensional case in similar fashion, using the four points that form a tetrahedron to show that if the flow vector is given by $\langle u, v, w \rangle$, then

$$(2.11) \qquad\qquad \frac{\partial u}{\partial x} + \frac{\partial v}{\partial y} + \frac{\partial w}{\partial z} = 0.$$

With a flow moving in the vertical direction that is subject to change over time, he extended his arguments to show that the pressure, $p = p(x, y, z, t)$, satisfies the following set of equations, rendered again in modern notation,

$$\frac{\partial p}{\partial x} = -2 \left(\frac{\partial u}{\partial x} u + \frac{\partial u}{\partial y} v + \frac{\partial u}{\partial z} w + \frac{\partial u}{\partial t} \right),$$

$$\frac{\partial p}{\partial y} = -2\left(\frac{\partial v}{\partial x}u + \frac{\partial v}{\partial y}v + \frac{\partial v}{\partial z}w + \frac{\partial v}{\partial t}\right),$$

$$\frac{\partial p}{\partial z} = -1 - 2\left(\frac{\partial w}{\partial x}u + \frac{\partial w}{\partial y}v + \frac{\partial w}{\partial z}w + \frac{\partial w}{\partial t}\right).$$

Euler would return to the problem of fluid dynamics many times over his career, often applying it to practical problems such as the most efficient shape for a ship's hull. His work with these partial differential equations constituted an important step toward the nineteenth-century derivation of the Navier-Stokes equations, discovered by Claude-Louis Navier (1785–1836) and George Gabriel Stokes (1819-1903) to describe the motion of viscous fluids. These form the foundation for the study of fluid mechanics.

2.10
The Vibrating String

Providing a mathematical model for physical phenomena, as Euler did for fluid dynamics, can provide great insight, both for explaining what is observed and for helping to shape the circumstances that will create desired outcomes. One clear example of this is the work on the mathematical model of a vibrating string undertaken in the first half of the eighteenth century. This has probably had a greater influence on our understanding of the physical world than any other mathematical model.

This story is worth telling for the insights it yields into how stringed instruments, such as violins or guitars, operate. It also provides the opening chapter for the story of the discovery of radio waves. From cell phones to garage door openers, everything requiring wireless connection in our modern world operates on a constant stream of these waves at multiple frequencies, yet their existence would never have been expected without the mathematical model of the phenomena of electricity and magnetism discovered by James Clerk Maxwell. So important is this collection of partial differential equations that Richard Feynman famously stated,

> From a long view of the history of mankind—seen from, say 10,000 years from now—there can be little doubt that the most significant

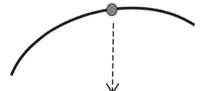

Figure 2.21. The restoring force on a plucked string is proportional to the second derivative of the curve at that point.

event of the 19th century will be judged as Maxwell's discovery of the laws of electrodynamics. The American Civil War will pale into provincial insignificance in comparison with this important scientific event of the same decade. (Feynman et al., 1964, vol. 2, section 1-6)

It was Brook Taylor in 1713 (the same Taylor for whom Taylor series are named), who first realized that the restoring force on any point of a plucked string depends on the second derivative of the curve formed by the deformed string at that point (see Figure 2.21). Although not a proof, this should make sense. If the string is locally linear (second derivative is zero), then there should be equal forces pulling it upward and downward, for a net force of zero. If the string is locally concave down, that point is being pulled down, and the greater the concavity (or more sharply it is bent), the greater one would expect the force to be.

Thirty-four years later, Jean le Rond d'Alembert (1717–1783) made the connection to the acceleration of that point on the string. As Newton had observed, the acceleration—the second derivative with respect to time— should be proportional to the force. If we let $h(x, t)$ denote the vertical displacement of a point on the string corresponding to position x and time t, then d'Alembert's realization amounts to the partial differential equation

(2.12)
$$\frac{\partial^2 h}{\partial x^2} = \frac{1}{c^2} \frac{\partial^2 h}{\partial t^2}.$$

To find a solution to this equation, we start with the special case in which h is the product of a function of x and a function of t,

$$h(x, t) = f(x)\, g(t).$$

Equation (2.12) becomes

$$f''(x) g(t) = \frac{1}{c^2} f(x) g''(t), \quad \text{or}$$

$$\frac{f''(x)}{f(x)} = \frac{1}{c^2} \frac{g''(t)}{g(t)},$$

assuming, for the moment, that neither f nor g is zero. But now the left side depends only on x and the right side depends only on t, so both of these must be constant, say equal to k^2:

$$f''(x) = k^2 f(x), \qquad g''(t) = k^2 c^2 g(t).$$

Now we have reduced the problem to a pair of simple differential equations whose solutions, up to multiplication by a constant, are the sine and cosine functions:

$$f(x) = \sin kx \quad \text{or} \quad \cos kx,$$

$$g(t) = \sin kct \quad \text{or} \quad \cos kct.$$

Whereas we found these solutions by assuming that f and g are not zero, we see that they work equally well when f or g is zero.

We now make two assumptions. The first is that the string is anchored at points $x = 0$ and $x = 1$, which forces f to be the sine function and k to be an integer multiple of π,

$$f(x) = \sin(m\pi x).$$

We next assume that at time 0, the string is stretched to its furthest extent, forcing g to be the cosine function, so that

$$g(t) = \cos(m\pi ct).$$

The constant c depends on the composition of the string and the tension it is under. When $m = 1$, we get the fundamental frequency: If time is measured in seconds, then it takes $2/c$ seconds to complete one cycle, so the frequency is $c/2$ cycles per second. Other values of m yield the *overtones*, frequencies of $mc/2$, all integer multiples of the fundamental frequency.

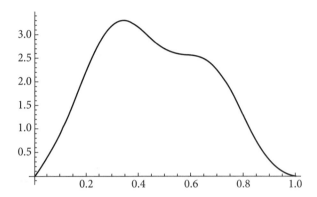

Figure 2.22. The graph of $y = 3 \sin(\pi x) + 0.5 \sin(2\pi x) - 0.3 \sin(5\pi x)$.

Not only is $f(x) g(t) = \sin(m\pi x) \cos(m\pi ct)$ a solution to equation (2.12), so is any linear combination of these functions, for example,

$$h(x, t) = 3 \sin(\pi x) \cos(\pi ct) + 0.5 \sin(2\pi x) \cos(2\pi ct)$$
$$- 0.3 \sin(5\pi x) \cos(5\pi ct).$$

Figure 2.22 shows the initial position $(t = 0)$ of the string that will vibrate with two overtones, one at $m = 2$ (an octave above the fundamental frequency), the other at two octaves and a third above the fundamental frequency.[41] Additional overtones can be created by changing how the string is plucked.

What happens if our string is not anchored, if we imagine an infinitely long taut string that is deformed over just a small stretch? In this case, the solution can be modeled by trigonometric functions of the form $h(x, t) = \cos(x - ct)$ traveling down the line at speed c. In the case of a two-dimensional wave—e.g., the disturbance to the surface of a still pond created by dropping a stone—the height of the water is a function of two position variables and time, $h(x, y, t)$, and the governing differential equation is

(2.13)
$$\frac{\partial^2 h}{\partial x^2} + \frac{\partial^2 h}{\partial y^2} = \frac{1}{c^2} \frac{\partial^2 h}{\partial t^2}.$$

In this case, we get a wave that travels out in circles at speed c.

2.11
The Power of Potentials

Pierre-Simon Laplace (1749–1827) was a protégé of d'Alembert who would become one of the leading European scientists of the late eighteenth to early nineteenth centuries. One of his greatest accomplishments was an exhaustive explanation of celestial mechanics—the movements of the planets, moons, and comets—in the five volumes of the *Traité de mécanique céleste* (Treatise on celestial mechanics), published over the years 1799 to 1825. Laplace built on Newton's *Principia* and simplified many of the arguments by relying on gravitational potential, rather than vectors of gravitational force.

The idea is related to potential energy. An object at a great height has more potential energy than one that is lower. It takes more energy to raise it to the higher position, and when we let it fall, it releases more energy. The advantage of potential energy is that it is a scalar, a single number. Rather than describing the force of gravity as a vector, showing the direction and magnitude of gravitational force, it is possible to encode gravitational attraction using a potential function which can be thought of as the height of a point on a hill. Just as water flows in the direction of steepest descent, with a velocity that is proportional to the steepness of that descent, gravitational attraction will always be in the direction of greatest change, with a strength that is proportional to the rate of change in that direction.

Given a potential function, often called a *potential field*, say $P(x, y, z)$, the vector describing the direction and magnitude of greatest change is given by what is called, appropriately, the *gradient*, denoted by grad P or ∇P. It is defined by

$$\nabla P = \left\langle \frac{\partial P}{\partial x}, \frac{\partial P}{\partial y}, \frac{\partial P}{\partial z} \right\rangle.$$

In a region where there are no bodies that either create or destroy gravitational force, the force lines will be incompressible. Combining the representation of the gradient with Euler's equation (2.11), we get what is known as Laplace's equation,

$$(2.14) \qquad \frac{\partial^2 P}{\partial x^2} + \frac{\partial^2 P}{\partial y^2} + \frac{\partial^2 P}{\partial z^2} = 0.$$

The expression on the left is known as the *Laplacian*, often written simply as $\nabla^2 P$.

If we are in a region where there *is* matter, that matter will produce a gravitational attraction on any other bodies with mass. If we think of gravitational lines of force as a flow emanating from a body with mass, then this flow is not incompressible at a point where mass exists, because it is actually being generated at that point. In this case, the Laplacian is no longer zero, but equal to the density of mass at that point, $\rho(x, y, z)$,

$$\nabla^2 P = \rho.$$

2.12
The Mathematics of Electricity and Magnetism

We started with operators and equations that applied to fluid flow and showed that, suitably interpreted, they can also be used to explain gravitational attraction. The breadth of applicability of these partial derivative equations is astonishing, something scientists came to realize in the nineteenth century as they began to unravel the mysteries of electricity and magnetism.

The word *electricity* is derived from the Latin, *electrum*, meaning amber. Before the nineteenth century, electricity was static electricity, the charge that could be obtained by rubbing a silk cloth against a piece of amber. It was a mystifying phenomenon that engaged the greatest scientific minds of the eighteenth century, most famously Philadelphia's own Benjamin Franklin. His experiments with lightning, Leiden jars (what today we call capacitors), and the electric jack (a primitive electric motor) brought him international fame. It was this fame that gave him entry to French society[42] and the diplomatic influence that would enable him to enlist the French into support of the Americans in the American Revolution.

Franklin was one of the first to realize that the creation of static electricity could be explained in terms of the flow of charge-carrying particles that exerted a force much like the force of gravity, except that it would repel rather than attract other particles. By the nineteenth century, thanks to the work of Daniel Bernoulli, Henry Cavendish, Charles Augustin Coulomb, and Carl Friedrich Gauss, this electrostatic force was described

Figure 2.23. James Clerk Maxwell.

by an electrostatic potential that satisfies exactly the same partial differential equation as gravitation attraction, but with ρ now representing charge density (and accompanied by a change of sign).

The early nineteenth century also witnessed the discovery of electrical current, the ability to generate a constant stream of electrons that move around a closed circuit. In 1820, Hans Christian Oersted (1777–1851), a Danish scientist, observed that an electrical circuit would deflect the needle of a compass when the circuit was closed. Word of this connection between electricity and magnetism spread quickly through Europe. By later the same year, the French scientists Jean-Baptiste Biot (1774–1862) and Félix Savart (1791–1841) had discovered the partial differential equation, known as Ampère's law, that encodes the relationship between a magnetic field and the electrical current that creates it.[43]

Electric current creates a magnetic field. As Michael Faraday (1791–1867) discovered in 1831, a moving magnet can also create an electric current. This, in fact, is how we generate electricity today, employing large spinning magnets known as dynamos. The partial differential equation that governs this interaction introduces a fourth variable, time.[44]

It was James Clerk Maxwell (1831–1879) who extended Ampère's law to include an electrical field that is changing over time and then realized the cohesiveness of all of these equations. In one of the most important scientific papers ever published, "A Dynamical Theory of the Electro-Magnetic Field," which appeared in the *Philosophical Transactions* of the Royal Society in 1865, he not only presented the equations that govern the interaction of changing electrical and magnetic fields, but he realized that everything could be simplified if expressed in terms of electro-magnetic potential.

Electro-magnetic potential is a strange idea. Gravitational potential is simple. It is a function from the three variables that describe position to the one variable that gives the value of the gravitational potential. Electro-magnetic potential depends on four variables, three of position and one of time, but it also produces four outputs, four separate functions that are each dependent on the four input variables. These four functions have no concrete meaning. Potential energy is a slippery concept. You cannot see it or touch it or directly experience it. It is really just a mathematical fiction that makes calculations much simpler. This is even truer for electro-magnetic potential. It has four components that appear to be nothing more than convenient way stations for mathematical calculations, much as complex numbers were long regarded as convenient fictions that facilitated the finding of roots of cubic and quartic polynomials.

But Maxwell discovered something else. Each component of the electro-magnetic potential, described as $\langle A_1, A_2, A_3, A_4 \rangle$, must satisfy the following partial differential equation:

$$(2.15) \qquad \frac{\partial^2 A_i}{\partial x^2} + \frac{\partial^2 A_i}{\partial y^2} + \frac{\partial^2 A_i}{\partial z^2} = \frac{1}{c^2} \frac{\partial^2 A_i}{\partial t^2},$$

where c is a constant that can be computed from the electrical and magnetic properties of the medium through which the electric and magnetic fields are interacting.

Surprisingly, equation (2.15) is a wave equation, an extension to three dimensions of d'Alembert's equation governing the vibrating string, equation (2.13). This strongly suggests that each component of the electro-magnetic potential vibrates, but also propagates as a wave traveling out in all directions in three-dimensional space at speed c.

Maxwell, measuring the electrical and magnetic properties of air, discovered that the speed of this propagation was, within the limits of experimental error, equal to the speed of light. He came to a remarkable conclusion: Not only is electro-magnetic potential something that actually exists, it vibrates through three-dimensional space, moving outward at the speed of light from the changing electro-magnetic field that created it.

His claim seemed incredible, but a few scientists believed there might be a reality behind this four-dimensional potential field. It was an intriguing idea. If it existed, then it might be possible to engineer a disturbance in the electro-magnetic potential that would spread at the speed of light and be detectable far away without any physical connection between the sender and receiver. In 1887, Heinrich Rudolf Hertz (1857–1894) managed to detect changes that he had produced in the electro-magnetic potential, and within a decade both Guglielmo Marconi (1874–1937) and Alexander S. Popov (1859–1906) had mastered the trick of converting Morse code to changes in the electro-magnetic potential that could be received and interpreted miles away.

Today we know these waves of electro-magnetic potential as radio waves. Wireless communication is based on them. Radio waves are intangible. They never would have been detected, much less exploited, without the mathematical model that predicted their existence.

The twentieth century saw growth in this power of differential equation models to suggest unexpected phenomena, including black holes, gravitational waves, and the fact that mass and energy are interchangeable via the relationship $E = mc^2$. These models employing partial differential equations sit behind virtually all of our modern technology.

Few of these differential equations can be solved using the standard repertoire of functions. Beginning with Newton and accelerating throughout the eighteenth and nineteenth centuries, scientists realized that solutions would require sophisticated use of infinite series. By the nineteenth century, understanding the intricacies of these strange summations would dominate much of the mathematical activity. Our next chapter will reveal some of their mysteries.

Chapter 3

SEQUENCES OF PARTIAL SUMS

One of the landmark years in the development of calculus was 1669 when Isaac Newton distributed his manuscript *On analysis by equations with an infinite number of terms*. When these infinite summations were combined with a clear understanding of the Fundamental Theorem of Integral Calculus, a shift of emphasis from geometry to dynamical systems, and the brilliant notation that Leibniz introduced, calculus was truly on its way. Clarity about what was fundamental and consistency in notation and argumentation enabled an army of scientists of the eighteenth century to achieve incredible successes in the understanding and application of calculus.

One must be careful when dealing with infinite series. As the Greeks realized, there is no such thing as an infinite summation. As stated in the preface, the term itself is an oxymoron, combining "infinite," meaning without end, and "summation," meaning the act of bringing to a conclusion. How do we bring to a conclusion something that never ends?

In Europe, we do not see significant work with infinite summations until the seventeenth century. The high point of seventeenth-century explorations, what today we call Taylor series, emerged from problems of polynomial interpolation. We will exhibit Euler's exuberant embrace of these series, describe growing concerns over questions of convergence, and conclude with an introduction to Fourier series, those infinite summations of trigonometric functions that would prove deeply problematic and motivate much of the development of analysis in the nineteenth century.

The problematic nature of infinite series is clear even for geometric series, the earliest of what today we know as infinite summations and the foundation for their study. The geometric series that begins with 1 and increases by a factor of x with each successive term is equal to $1/(1-x)$

in some cases, but not in others. Thus,

$$(3.1) \qquad 1 + \frac{2}{3} + \frac{4}{9} + \frac{8}{27} + \cdots + \frac{2^n}{3^n} + \cdots = \frac{1}{1 - 2/3} = 3, \text{but}$$

$$(3.2) \qquad 1 + \frac{3}{2} + \frac{9}{4} + \frac{27}{8} + \cdots + \frac{3^n}{2^n} + \cdots \neq \frac{1}{1 - 3/2} = -2.$$

Archimedes proved that the area of a parabolic segment (the area between the arc of a parabola and a straight line that cuts it twice) is 4/3rds the area of the largest triangle that can be inscribed in this region. He proceeded by inserting this triangle and showing that two more triangles of total area 1/4 that of the original triangle can be inserted into the two segments that remain. After the kth iteration, we have built up an area of

$$\left(1 + \frac{1}{4} + \frac{1}{4^2} + \cdots + \frac{1}{4^k} \right) \text{ times the area of the triangle.}$$

Just as with his proof for the area of a circle, he observed that with each insertion, we have accounted for more than half of the remaining area. Now noting that

$$\frac{4}{3} - \left(1 + \frac{1}{4} + \frac{1}{4^2} + \cdots + \frac{1}{4^k} \right) = \frac{1}{3 \cdot 4^{k+1}},$$

he concluded that the true area is neither less than nor greater than 4/3rds the area of the original triangle.

In modern language, he had proven that the infinite series

$$1 + \frac{1}{4} + \frac{1}{4^2} + \cdots + \frac{1}{4^k} + \cdots$$

is equal to 4/3. What we mean by this today is precisely what Archimedes would have accepted: Given any number less than 4/3, the partial sums will eventually be larger than that number, and given any number greater than 4/3, they will eventually (in this case, always) be less than that number.

Augustin-Louis Cauchy (1789–1857) codified this interpretation of an infinite summation. The distinction between equations (3.1) and (3.2) is apparent if we look at finite sums,

$$(3.3) \quad 1 + \frac{2}{3} + \frac{4}{9} + \frac{8}{27} + \cdots + \frac{2^n}{3^n} = \frac{1 - 2^{n+1}/3^{n+1}}{1 - 2/3} = 3 - \frac{2^{n+1}}{3^n}, \text{ while}$$

$$(3.4) \quad 1 + \frac{3}{2} + \frac{9}{4} + \frac{27}{8} + \cdots + \frac{3^n}{2^n} = \frac{1 - 3^{n+1}/2^{n+1}}{1 - 3/2} = -2 + \frac{3^{n+1}}{2^n}.$$

In equation (3.3), the distance of the partial sum from 3 can be brought as close as we wish to zero by taking n sufficiently large. In equation (3.4), the distance from -2 gets larger as n increases. When we speak of an infinite series, what we really mean is a sequence of partial sums. We will see that this is the way Mengoli, Leibniz, and Lagrange thought about these series.

3.1
Series in the Seventeenth Century

It appears that François Viète, in 1593, was the first to describe a summation as continuing "ad infinitum." Pietro Mengoli (1626–1686), a student of Cavalieri and his successor as professor at the University of Bologna, was one of the first European philosophers to explore such sums. In *Novæ quadraturæ arithmeticæ, sue de additione fractionum* (New arithmetic of areas, and the addition of fractions), published in 1650, he thought of infinite sums as the values that are approached by the sequence of partial sums and based his study of infinite series of positive terms on two important axioms, or assumptions, which I recast into modern language:

(1) If the value of the series is infinite, then given any positive number, eventually the partial sums will be greater than that value.
(2) If the value of the series is finite, then any rearrangement of the series gives the same value.

Today, we would derive these properties from the definition of the limit of the sequence of partial sums. Mengoli used these assumptions to derive several properties of infinite series. Keeping in mind that he assumed that all terms are positive, he argued that if the partial sums are bounded, then the series must converge. He also demonstrated that if a convergent series has value S and A is *any* value less than S, then eventually the partial sums will be greater than A.

Mengoli used partial fraction decomposition to find the values of several series, starting with the series that Leibniz would sum early in his career (equation (2.8) in section 2.6). Mengoli also summed other series for which he could use partial fraction decompositions. For example, because

$$\frac{1}{n(n+2)} = \frac{1}{2}\left(\frac{1}{n} - \frac{1}{n+2}\right),$$

we see that

$$\sum_{n=1}^{m} \frac{1}{n(n+2)} = \frac{1}{1\cdot 3} + \frac{1}{2\cdot 4} + \frac{1}{3\cdot 5} + \cdots + \frac{1}{(m-1)(m+1)} + \frac{1}{m(m+2)}$$

$$= \frac{1}{2}\left(1 - \frac{1}{3} + \frac{1}{2} - \frac{1}{4} + \frac{1}{3} - \frac{1}{5} + \cdots\right.$$

$$\left. + \frac{1}{m-1} - \frac{1}{m+1} + \frac{1}{m} - \frac{1}{m+2}\right)$$

$$= \frac{1}{2}\left(1 + \frac{1}{2} - \frac{1}{m+1} - \frac{1}{m+2}\right).$$

The series converges to 3/4. Other sums that he evaluated included

$$\sum_{n=1}^{\infty} \frac{1}{n(n+3)} = \frac{11}{18},$$

$$\sum_{n=1}^{\infty} \frac{1}{n(n+1)(n+2)} = \frac{1}{4},$$

$$\sum_{n=1}^{\infty} \frac{1}{(2n-1)(2n+1)(2n+3)} = \frac{1}{12}.$$

Mengoli also gave the following proof of the divergence of the harmonic series,[1]

$$1 + \frac{1}{2} + \frac{1}{3} + \frac{1}{4} + \cdots.$$

We assume that it has a finite value. Since

$$\frac{1}{3n-1} + \frac{1}{3n} + \frac{1}{3n+1} = \frac{27n^2-1}{27n^3-3n} > \frac{1}{n},$$

the harmonic series, which can be written as

$$1 + \left(\frac{1}{2} + \frac{1}{3} + \frac{1}{4}\right) + \left(\frac{1}{5} + \frac{1}{6} + \frac{1}{7}\right) + \left(\frac{1}{8} + \frac{1}{9} + \frac{1}{10}\right) + \cdots,$$

has a value greater than

$$1 + 1 + \frac{1}{2} + \frac{1}{3} + \frac{1}{4} + \cdots.$$

This is one more than its value, an impossibility. He also observed that $\sum 1/n^2$ must converge (its summands are bounded above by $\frac{2}{n(n+1)}$). Finding the actual value of this series would take 85 years. It would constitute one of the early achievements of Leonhard Euler.

By 1693, Leibniz realized the potential of power series, infinite summations of the form $\sum c_k x^k$, to solve differential equations. In *Supplementum geometræ practicæ*[2] he showed how to use the "method of indeterminate coefficients" to find a solution to the differential equation

$$dy = \frac{a\,dx}{a+x}.$$

Dividing by dx and multiplying through by $a + x$, he obtained

(3.5)
$$0 = a\frac{dy}{dx} + x\frac{dy}{dx} - a.$$

He now assumed that the solution is zero when $x = 0$ and wrote it as[3]

$$y = c_1 x + c_2 x^2 + c_3 x^3 + c_4 x^4 + \cdots.$$

Differentiating yields

$$\frac{dy}{dx} = c_1 + 2c_2 x + 3c_3 x^2 + 4c_4 x^3 + \cdots.$$

We substitute the series for dy/dx in equation (3.5) and collect the coefficients of the same powers of x,

$$
\begin{aligned}
0 = {} & ac_1 + 2ac_2x + 3ac_3x^2 + 4ac_4x^3 + 5c_5x^4 \cdots \\
& + c_1x + 2c_2x^2 + 3c_3x^3 + 4c_4x^4 + \cdots - a \\
= {} & a(c_1 - 1) + (2ac_2 + c_1)x + (3ac_3 + 2c_2)x^2 + (4ac_4 + 3c_3)x^3 \\
& + (5ac_5 + 4c_4)x^4 + \cdots .
\end{aligned}
$$

Since this series equals 0, each coefficient must be zero, yielding

$$a(c_1 - 1) = 0 \Longrightarrow c_1 = 1,$$

$$2ac_2 + c_1 = 0 \Longrightarrow c_2 = -\frac{1}{2a},$$

$$3ac_3 + 2c_2 = 0 \Longrightarrow c_3 = \frac{1}{3a^2},$$

$$4ac_4 + 3c_3 = 0 \Longrightarrow c_4 = -\frac{1}{4a^3},$$

$$5ac_5 + 4c_4 = 0 \Longrightarrow c_4 = \frac{1}{5a^4},$$

$$\vdots$$

The solution to this differential equation with $y = 0$ when $x = 0$ is

$$y = x - \frac{x^2}{2a} + \frac{x^3}{3a^2} - \frac{x^4}{4a^3} + \frac{x^5}{5a^4} - \cdots .$$

Recognizing that $a \ln(1 + x/a)$ is a solution to the differential equation and that it is 0 when $x = 0$, Leibniz had proven that

$$a \ln(1 + x/a) = x - \frac{x^2}{2a} + \frac{x^3}{3a^2} - \frac{x^4}{4a^3} + \frac{x^5}{5a^4} - \cdots .$$

Leibniz also rediscovered the identity

$$1 - \frac{1}{3} + \frac{1}{5} - \frac{1}{7} + \cdots = \frac{\pi}{4}.$$

Today, we are so accustomed to equalities that involve infinite series that this statement seems unexceptional. But this is not an equality in the ordinary sense because the series is not a summation in the ordinary sense. Leibniz was clearly uneasy about so boldly asserting this as an equality. In a paper published in 1682 in *Acta Eruditorum*, he began by establishing convergence of the sequence of partial sums. In what appears to be the first application of the alternating series test, he observed that the first term of this series is within 1/3 of $\pi/4$, the first two terms together are within 1/5, the first three terms within 1/7, and so on. Thus the difference between the values of the partial sums and $\pi/4$ becomes less than any given quantity. He then argued that one should be allowed to consider the entire infinite series as an entity in its own right. If we do, then its value can only be $\pi/4$.

3.2
Taylor Series

As we have seen, work on areas and volumes in the seventeenth century relied on infinite summations, which were treated with more or less rigor depending on the author and the degree of belief that one must hew closely to the Archimedean approach. Today, the study of infinite series in first-year calculus is dominated by the study of what the English-speaking world refers to as Taylor series, named for Brook Taylor who popularized them in the early eighteenth century.

The construction of Taylor series was an open secret freely shared among the late seventeenth-century developers of calculus, including James Gregory, Isaac Newton, Gottfried Leibniz, and Jacob and Johann Bernoulli. When Taylor finally published the general form of these series with coefficients determined by the derivatives of the function to be represented, he developed it as a corollary of Newton's interpolation formula. This is likely how many others found it.

Both James Gregory and Isaac Newton began with the question: How can we find the *interpolating polynomial*[4] of degree n that passes through a set of $n+1$ points (see Figure 3.1)?

If we just have two points, say $(x_0, y_0) = (0, 1)$ and $(x_1, y_1) = (2, 4)$, they determine a straight line given by the equation

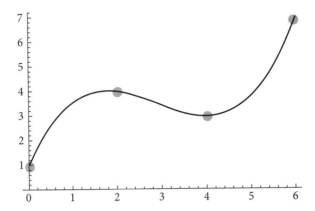

Figure 3.1. The graph of the cubic equation that passes through the points $(0, 1)$, $(2, 4)$, $(4, 3)$, and $(6, 7)$: $y = 1 + 4x - \frac{13}{8}x^2 + \frac{3}{16}x^3$.

$$y = y_0 + \frac{y_1 - y_0}{x_1 - x_0}(x - x_0) = 1 + \frac{3}{2}(x - 0) = 1 + \frac{3}{2}x.$$

For convenience, we denote

$$y_1 - y_0 = \Delta y_0, \qquad x_1 - x_0 = \Delta x.$$

We now can write the general equation of the straight line as

$$y = y_0 + \frac{\Delta y_0}{\Delta x}(x - x_0).$$

What if we have three points with equally spaced x-coordinates, $(x_0, y_0) = (0, 1)$, $(x_0 + \Delta x, y_1) = (2, 4)$, and $(x_0 + 2\Delta x, y_2) = (4, 3)$? Three points uniquely determine a quadratic equation. To find it, we begin by looking at the change in y between the first and second and the second and third points,

$$\Delta y_0 = 4 - 1 = 3, \qquad \Delta y_1 = y_2 - y_1 = 3 - 4 - = -1.$$

Now look at the change in the changes,

$$\Delta y_1 - \Delta y_0 = (y_2 - y_1) - (y_1 - y_0) = -1 - 3 = -4.$$

We write this as

$$\Delta \, \Delta y_0 = \Delta^2 y_0 = -4.$$

The quadratic equation that passes through our three points is

(3.6)
$$y = y_0 + \frac{\Delta y_0}{\Delta x}(x - x_0) + \frac{\Delta^2 y_0}{2(\Delta x)^2}(x - x_0)(x - x_1).$$

It is easy to check that (x_0, y_0) and (x_1, y_1) satisfy this equation. When $x = x_2 = x_0 + 2\Delta x$, the right side of our equation is

$$y_0 + \frac{y_1 - y_0}{\Delta x}(2\Delta x) + \frac{y_2 - 2y_1 + y_0}{2\Delta x^2}(\Delta x)(2\Delta x)$$

$$= y_0 + 2y_1 - 2y_0 + y_2 - 2y_1 + y_0 = y_2.$$

The quadratic equation that passes through the three given points is

$$y = 1 + \frac{3}{2}(x - 0) + \frac{-4}{2 \cdot 2^2}(x - 0)(x - 2)$$

$$= 1 + \frac{3}{2}x - \frac{1}{2}x(x - 2)$$

$$= 1 + \frac{5}{2}x - \frac{1}{2}x^2.$$

This method of quadratic interpolation is old. It was known to the Indian astronomers of the first millennium of the Common Era.

What if we have more points through which we must weave our polynomial? We can take differences of differences. We define the kth difference inductively,

$$\Delta^k y_n = \Delta \left(\Delta^{k-1} y_n \right) = \Delta^{k-1} y_{n+1} - \Delta^{k-1} y_n.$$

Both Gregory and Newton discovered how to use these differences to construct the polynomial of degree n that goes through the points $(x_0, y_0), (x_1, y_1), \ldots, (x_n, y_n)$. To keep things simple, we will assume that the x-values are equally spaced,

$$x_1 - x_0 = x_2 - x_1 = \cdots = x_n - x_{n-1} = \Delta x.$$

We take our y-values and consider their differences, the differences of the differences, and so on. As we will see, in each successive row we can stop one difference short of the previous row:

$$
\begin{array}{cccccccc}
y_0 & \Delta y_0 & \Delta^2 y_0 & \cdots & \Delta^{n-2} y_0 & \Delta^{n-1} y_0 & \Delta^n y_0 \\
y_1 & \Delta y_1 & \Delta^2 y_1 & \cdots & \Delta^{n-2} y_1 & \Delta^{n-1} y_1 \\
y_2 & \Delta y_2 & \Delta^2 y_2 & \cdots & \Delta^{n-2} y_2 \\
\vdots & \vdots & \vdots \\
y_{n-1} & \Delta y_{n-1} \\
y_n
\end{array}
$$

We used all of the values in the first column to construct the differences in the first row. What Gregory and Newton realized was that they could use the first row to reconstruct the first column. We built the first row by taking differences. To reverse the process, we take sums:

$$y_1 = y_0 + \Delta y_0, \quad \Delta y_1 = \Delta y_0 + \Delta^2 y_0, \quad \cdots \quad , \Delta^{n-1} y_1 = \Delta^{n-1} y_0 + \Delta^n y_0.$$

At the next row, things get interesting:

$$y_2 = y_1 + \Delta y_1 = (y_0 + \Delta y_0) + (\Delta y_0 + \Delta^2 y_0) = y_0 + 2\Delta y_0 + \Delta^2 y_0.$$

In general,

$$
\begin{aligned}
\Delta^j y_2 &= \Delta^j y_1 + \Delta^{j+1} y_1 \\
&= (\Delta^j y_0 + \Delta^{j+1} y_0) + (\Delta^{j+1} y_0 + \Delta^{j+2} y_0) \\
&= \Delta^j y_0 + 2\Delta^{j+1} y_0 + \Delta^{j+2} y_0.
\end{aligned}
$$

For the fourth row, we see that

$$
\begin{aligned}
y_3 &= y_2 + \Delta y_2 \\
&= (y_0 + 2\Delta y_0 + \Delta^2 y_0) + (\Delta y_0 + 2\Delta^2 y_0 + \Delta^3 y_0) \\
&= y_0 + 3\Delta y_0 + 3\Delta^2 y_0 + \Delta^3 y_0.
\end{aligned}
$$

In general,

$$\Delta^j y_3 = \Delta^j y_0 + 3\Delta^{j+1} y_0 + 3\Delta^{j+1} y_0 + \Delta^{j+3} y_0.$$

The binomial coefficients emerge. They appear because each new term arises from the addition of the term directly above and the term above and to the right. It follows that

$$y_k = y_0 + \binom{k}{1}\Delta y_0 + \binom{k}{2}\Delta^2 y_0 + \cdots + \binom{k}{k}\Delta^k y_0.$$

To find the interpolating polynomial for our $n+1$ points, we only need to find a polynomial that takes on the appropriate y_k value at x_0, $x_1 = x_0 + \Delta x$, $x_2 = x_0 + 2\Delta x$, $\ldots$, $x_n = x_0 + n\Delta x$. This polynomial follows the pattern given in equation (3.6), namely,

$$(3.7) \qquad p_n(x) = y_0 + \sum_{j=1}^{n} \frac{\Delta^j y_0}{j!\,(\Delta x)^j}(x - x_0)(x - x_1)(x - x_2)\cdots(x - x_{j-1}).$$

To prove this, we first observe that this polynomial equals y_0 when $x = x_0$ because only this initial constant is not multiplied by $x_0 - x_0 = 0$. If $x = x_k$, $1 \leq k \leq n$, then only the first k terms of the summand survive, and the polynomial takes the value

$$p_n(x_k) = y_0 + \sum_{j=1}^{k} \frac{\Delta^j y_0}{j!\,(\Delta x)^j}(x_k - x_0)(x_k - x_1)(x_k - x_2)\cdots(x_k - x_{j-1})$$

$$= y_0 + \sum_{j=1}^{k} \frac{\Delta^j y_0}{j!\,(\Delta x)^j}\, k\Delta x \cdot (k-1)\Delta x \cdot (k-2)\Delta x \cdots (k-j+1)\Delta x$$

$$= y_0 + \sum_{j=1}^{k} \frac{k(k-1)(k-2)\cdots(k-j+1)}{j!}\,\Delta^j y_0$$

$$= y_0 + \binom{k}{1}\Delta y_0 + \binom{k}{2}\Delta^2 y_0 + \cdots + \binom{k}{k}\Delta^k y_0$$

$$= y_k.$$

Newton recognized the importance of this interpolating polynomial for approximating integrals. If we know that $y_0 = f(x_0)$ and $y_1 = f(x_1)$, then the integral of f from x_0 to x_1 can be approximated by integrating the linear function through the two known points:

$$\int_{x_0}^{x_1} f(x)\, dx \approx \int_{x_0}^{x_1} \left(y_0 + \frac{y_1 - y_0}{x_1 - x_0}(x - x_0) \right) dx$$

$$= y_0(x_1 - x_0) + \frac{y_1 - y_0}{2(x_1 - x_0)}(x_1 - x_0)^2$$

$$= (x_1 - x_0)\frac{y_1 + y_0}{2}.$$

This is the trapezoidal rule.

What if we have three equally spaced points: (x_0, y_0), $(x_0 + \Delta x, y_1)$, $(x_0 + 2\Delta x, y_2)$? Integrating the quadratic polynomial through these points gives us

$$\int_{x_0}^{x_2} f(x)\, dx \approx \int_{x_0}^{x_2} \left(y_0 + \frac{y_1 - y_0}{\Delta x}(x - x_0) + \frac{y_2 - 2y_1 + y_0}{2\Delta x^2}(x - x_0)(x - x_1) \right) dx$$

$$= y_0 \cdot 2\Delta x + \frac{y_1 - y_0}{\Delta x} \cdot 2\Delta x^2 + \frac{y_2 - 2y_1 + y_0}{2\,\Delta x^2} \cdot \frac{2}{3}\Delta x^3$$

$$= \frac{\Delta x}{3}(y_0 + 4y_1 + y_2).$$

This is known as Simpson's rule, named for Thomas Simpson (1710–1761) who rediscovered it in 1743.

Newton also gave the formula for the next case, using the polynomial of degree three,

$$\int_{x_0}^{x_3} f(x)\, dx \approx \frac{3\Delta x}{8}(y_0 + 3y_1 + 3y_2 + y_3),$$

known as the Newton-Cotes three-eighths rule.

Following Brook Taylor, if we take n to be unlimited (we know the value of the function at infinitely many points) and we let Δx approach 0, then the fraction $\Delta y_0 / \Delta x$ becomes dy/dx (evaluated at x_0). The fraction

$$\frac{\Delta^2 y_0}{\Delta x^2} = \frac{\Delta}{\Delta x}\left(\frac{\Delta y_0}{\Delta x} \right)$$

becomes

$$\frac{d}{dx}\left(\frac{dy}{dx}\right) = \frac{d^2y}{dx^2}.$$

In general $\Delta^k y_0 / \Delta x^k$ becomes $d^k y / dx^k$. At the same time, all of the points $x_1 = x_0 + \Delta x, x_2 = x_0 + 2\Delta x, \ldots$ slide over to become just x_0. Our polynomial has become the Taylor series,

$$p_\infty(x) = y_0 + \frac{dy}{dx}(x - x_0) + \frac{1}{2!} \cdot \frac{d^2y}{dx^2}(x - x_0)^2 + \frac{1}{3!} \cdot \frac{d^3y}{dx^3}(x - x_0)^3 + \cdots,$$

where all of the derivatives are evaluated at x_0.

Colin Maclaurin (1698–1746) knew of, and in fact referenced, Taylor's work when he highlighted the special case of Taylor's series that is centered at the origin, what today are called Maclaurin series. This was published in his 1742 explanation of Newton's methods, *Treatise of fluxions*.

3.3
Euler's Influence

In 1689, Jacob Bernoulli published *Tractatus de seriebus infinitis* (Treatise on infinite series). He included Mengoli's problem, to find the exact value of the convergent series

$$\sum_{n=1}^{\infty} \frac{1}{n^2}.$$

Now known as "the Basel Problem" and accompanied by a suspicion that the value involved π, this became one of the great challenges of the era. By 1729, Euler had refined methods for approximating infinite series to the point where he knew the first seven digits: 1.644934. This led him to suspect that the series equals $\pi^2/6 = 1.6449340668\ldots$. In 1734, he proved it, thus establishing his reputation. His methods here and in other work on series were not rigorous by today's standards. His representation of $(\sin x)/x$ as an infinite product would not be fully justified until 1876.[5]

Euler's solution to the Basel problem arose from his work on polynomials. He recognized that if the roots of a polynomial, $r_1, r_2, \ldots, r_k$, are known, they uniquely determine that polynomial up to a constant,

$$p_k(x) = c(x - r_1)(x - r_2) \cdots (x - r_k).$$

If 0 is not a root and we normalize our polynomial so that $p_k(0) = 1$, then that polynomial can be expressed as

$$(3.8) \qquad p_k(x) = \left(1 - \frac{x}{r_1}\right)\left(1 - \frac{x}{r_2}\right) \cdots \left(1 - \frac{x}{r_k}\right).$$

Euler now made the leap to the function $\frac{\sin x}{x}$, which can be taken to be 1 at $x = 0$.[6] This function has roots at all nonzero integer multiples of π. Assuming that equation (3.8) would continue to be valid when k is infinite, he boldly asserted that

$$\frac{\sin x}{x} = \left(1 - \frac{x}{\pi}\right)\left(1 - \frac{x}{-\pi}\right)\left(1 - \frac{x}{2\pi}\right)\left(1 - \frac{x}{-2\pi}\right) \cdots .$$

He combined pairs of products,

$$\left(1 - \frac{x}{k\pi}\right)\left(1 - \frac{x}{-k\pi}\right) = \left(1 - \frac{x^2}{k^2\pi^2}\right),$$

and expanded the resulting product, collecting terms that multiply each power of x,

$$\frac{\sin x}{x} = \left(1 - \frac{x^2}{\pi^2}\right)\left(1 - \frac{x^2}{2^2\pi^2}\right)\left(1 - \frac{x^2}{3^2\pi^2}\right) \cdots$$

$$= 1 - \left(1 + \frac{1}{2^2} + \frac{1}{3^2} + \cdots\right)\frac{x^2}{\pi^2} + \left(\sum_{1 \le i < j < \infty} \frac{1}{i^2 j^2}\right)\frac{x^4}{\pi^4} - \cdots .$$

Given the Taylor series

$$\frac{\sin x}{x} = 1 - \frac{x^2}{3!} + \frac{x^4}{5!} - \frac{x^6}{7!} + \cdots ,$$

he equated the coefficients of x^2 to get[7]

$$-\left(1 + \frac{1}{2^2} + \frac{1}{3^2} + \cdots\right)\frac{1}{\pi^2} = -\frac{1}{3!},$$

$$1 + \frac{1}{2^2} + \frac{1}{3^2} + \cdots = \frac{\pi^2}{6}.$$

In his two-volume work, *Introduction to analysis of the infinite*, published in 1748, Leonhard Euler produced what Boyer has called "the most influential mathematics textbook" of modern times. It was this text that raised functions and power series to their role as fundamental concepts upon which modern calculus is built.

As we have seen, both Newton and Leibniz understood the importance of power series and moved calculus away from purely geometric considerations toward the study of functional relationships. But neither the central importance of power series nor the usefulness of constructing calculus around the concept of function was made as explicit as Euler was to make it in his *Introduction*. A curious feature of Euler's *Introduction to analysis of the infinite* is that it contains not a single derivative or integral. Euler conceived it as a precalculus text, but one with infinitesimals and infinities.

The first three chapters of Euler's *Introduction* were devoted to explaining the concept of *function* and exploring functional transformations. He then found infinite series expansions of transcendental functions, which he accomplished by algebraic means, arguing from analogy. Euler carried the Bernoullis' acceptance of infinitesimals and infinities to a dangerous extreme, yet he found his way through this minefield with a dexterity that is often breathtaking. I am including his derivation of the power series expansions of the exponential and logarithmic functions, not because they are examples worthy of emulation, but because they exhibit a joy in the exploration of infinite series that I wish today's students could share.

Euler began with an arbitrary base, $a > 1$, for his exponential function, a^x. When $x = 0$, this is 1, when $x > 0$ it is larger than 1, when $x < 0$ it is less than one. He now let ω denote a fixed infinitely small number (positive or negative) and wrote

$$a^{\omega} = 1 + k\omega,$$

where k will be some positive quantity that depends on a.

Because $a^{\omega} = 1 + k\omega$, we also have that, for any number j,

$$a^{j\omega} = (1 + k\omega)^j = 1 + \frac{j}{1}k\omega + \frac{j(j-1)}{1 \cdot 2}k^2\omega^2 + \frac{j(j-1)(j-2)}{1 \cdot 2 \cdot 3}k^3\omega^3 + \cdots.$$

Now set $j\omega = z$, so that $\omega = z/j$. Then

$$a^z = 1 + \frac{1}{1}kz + \frac{1(j-1)}{1 \cdot 2j}k^2z^2 + \frac{1(j-1)(j-2)}{1 \cdot 2 \cdot 3j^2}k^3z^3 + \cdots.$$

If z is finite, then j must be infinite, but then $(j-1)/j = (j-2)/j = (j-3)/j = \cdots = 1$. We thus have our power series expansion of a^z,

$$a^z = 1 + kz + \frac{k^2 z^2}{1 \cdot 2} + \frac{k^3 z^3}{1 \cdot 2 \cdot 3} + \cdots .$$

Setting $z = 1$ yields a as a function of k,

$$a = 1 + k + \frac{k^2}{1 \cdot 2} + \frac{k^3}{1 \cdot 2 \cdot 3} + \cdots .$$

Euler next tackled the power series for the logarithm. Because $a^\omega = 1 + k\omega$, we know that $\omega = \log_a(1 + k\omega)$, and therefore $j\omega = \log_a(1 + k\omega)^j$. Because for positive values of j, $(1 + k\omega)^j$ will be larger than 1, he wrote $(1 + k\omega)^j = 1 + x$. If we take x to be a finite number, then j must be infinite. A little rearranging led him to

$$k\omega = -1 + (1 + x)^{1/j}$$

$$= -1 + 1 + \frac{\left(\frac{1}{j}\right)}{1}x + \frac{\left(\frac{1}{j}\right)\left(\frac{1}{j} - 1\right)}{1 \cdot 2}x^2 + \frac{\left(\frac{1}{j}\right)\left(\frac{1}{j} - 1\right)\left(\frac{1}{j} - 2\right)}{1 \cdot 2 \cdot 3}x^3 + \cdots$$

$$= \frac{1}{j}x + \frac{1(1 - j)}{1 \cdot 2j^2}x^2 + \frac{1(1 - j)(1 - 2j)}{1 \cdot 2 \cdot 3j^3}x^3 + \cdots .$$

Because j is infinite, $(1 - j)/j = -1$, $(1 - 2j)/j = -2$, $(1 - 3j)/j = -3$, and so on. Euler had demonstrated that

$$jk\omega = x - \frac{x^2}{2} + \frac{x^3}{3} - \frac{x^4}{4} + \cdots .$$

However, we know that $jk\omega = k\log_a(1 + k\omega)^j = k\log_a(1 + x)$. It follows that

$$\log_a(1 + x) = \frac{1}{k}\left(x - \frac{x^2}{2} + \frac{x^3}{3} - \frac{x^4}{4} + \cdots\right).$$

We now choose the value of a so that $k = 1$. This value is

$$a = 1 + 1 + \frac{1}{1 \cdot 2} + \frac{1}{1 \cdot 2 \cdot 3} + \cdots = 2.71828182845904523536028 \ldots .$$

Euler actually gave all of these digits, then announced that he would call this constant e.

It is now commonplace to introduce power series as a tool for approximating the values of transcendental functions. That is an important use, but not their real significance. Some instructors even go so far as to suggest that they explain how calculators determine the values of trigonometric, exponential, and logarithmic functions (not true). Their real usefulness is the ease with which they can be differentiated and integrated.

Once we know that $e^x = 1 + x + x^2/2! + x^3/3! + \cdots$, it is easy to see that the derivative of e^x is once again e^x. From the power series expansion of the natural logarithm of $1 + x$, one easily computes its derivative,

$$\frac{d}{dx}\ln(1+x) = \frac{d}{dx}\left(x - \frac{x^2}{2} + \frac{x^3}{3} - \frac{x^4}{4} + \cdots\right)$$
$$= 1 - x + x^2 + x^3 - x^4 + \cdots$$
$$= \frac{1}{1+x},$$

because this is a convergent geometric series for $-1 < x < 1$. Integration is no harder. If we want to integrate e^{x^2}, we begin with its power series expansion, then integrate each term:

$$\int_0^x e^{t^2}\,dt = \int_0^x \left(1 + t^2 + \frac{t^4}{2!} + \frac{t^6}{3!} + \cdots\right)dt$$
$$= x + \frac{x^3}{3} + \frac{x^5}{2!\cdot 5} + \frac{x^7}{3!\cdot 7} + \cdots.$$

The fact that this series does not have a representation in terms of common transcendental functions in no way detracts from the fact that this *is* a perfectly acceptable solution.

Power series also lead us to immensely useful insights. Euler used the power series for the exponential function to justify that $e^{ix} = \cos x + i\sin x$,

$$(3.9) \qquad e^{ix} = 1 + ix + \frac{i^2 x^2}{2!} + \frac{i^3 x^3}{3!} + \frac{i^4 x^4}{4!} + \frac{i^5 x^5}{5!} + \cdots$$
$$= 1 + ix - \frac{x^2}{2!} - \frac{ix^3}{3!} + \frac{x^4}{4!} + \frac{ix^5}{5!} + \cdots$$

$$= \left(1 - \frac{x^2}{2!} + \frac{x^4}{4!} - \cdots\right) + i\left(x - \frac{x^3}{3!} + \frac{x^5}{5!} - \cdots\right)$$

$$= \cos x + i \sin x.$$

3.4
D'Alembert and the Problem of Convergence

Although Euler understood the importance of convergence of an infinite series, he refused to be limited by it. Thus, he asserted that $1 + x + x^2 + x^3 + \cdots$ always equals $1/(1-x)$, recognizing that the series only converges to that value when $|x| < 1$, but allowing this expression to stand as a defining value for this series when $|x| \geq 1$, $x \neq 1$. In his paper "On divergent series" of 1760, he remarked that the value of a "sum" depends on how we choose to define this term.

> [If] the sum of a series is said to be that quantity to which it is brought closer as more terms of the series are taken [then this] has relevance only for convergent series. On the other hand, as series in analysis arise from the expansion of fractions or irrational quantities or even of transcendentals, it will in turn be permissible in calculation to substitute in place of such series that quantity out of whose development it is produced. (Barbeau and Leah, 1976, p. 144)

One of the first to directly tackle the problem of the convergence of an infinite series was Jean le Rond d'Alembert, whom we met in the previous chapter for his work on the mathematical model of a vibrating string. In 1768 he considered the problem of the convergence of the binomial series,

(3.10)
$$(1+x)^m = 1 + mx + \frac{m(m-1)}{2!}x^2 + \cdots + \frac{m(m-1)\cdots(m-j+1)}{j!}x^j + \cdots.$$

In this paper, he became the first to describe the *ratio test*.

Jean le Rond d'Alembert (Figure 3.2) is one of my favorite philosophers of mathematics. Although he was born after 1700, "philosopher" describes him more accurately than "scientist," for his interests were far-ranging. Born the illegitimate offspring of a liaison between Claudine

Figure 3.2. Jean le Rond d'Alembert. Pastel by Maurice Quentin de La Tour.

Guérin de Tencin and the chevalier Louis-Camus Destouches, he was left on the steps of the church of St. Jean le Rond in Paris, from which he was given the name Jean le Rond. The designation Le Rond or The Round refers to the shape of the church rather than that of Saint John. The surname d'Alembert was his own invention. His father, the chevalier Destouches, saw that he was properly educated, and d'Alembert became one of the great philosophers of the Enlightenment. In the 1740s, he joined Denis Diderot (1713–1784) in the ambitious project of putting all of human knowledge in one place, the *Encyclopédie, ou dictionnaire raisonné des sciences, des art et des métiers* (Encyclopedia, or descriptive dictionary of the sciences, arts, and crafts), which would run to 28 volumes and to which d'Alembert contributed over 1300 articles.

D'Alembert began his study of convergence by observing that the summand with x^j is obtained from the summand with x^{j-1} by multiplying the latter by

$$\frac{m-j+1}{j}x = \left(-1 + \frac{m+1}{j}\right)x,$$

equivalent to taking the ratio of consecutive terms. As j gets larger, the quantity $\left(-1 + \frac{m+1}{j}\right)$ gets closer to -1, so that if $|x| > 1$, then

$$\left|\left(-1+\frac{m+1}{j}\right)x\right| > 1$$

when j is sufficiently large. Because the summands are getting larger, the series cannot possibly converge.

How large does j need to be before the ratio of summands is larger than 1? If m and x are positive and $j > m$, then we need

$$\frac{j-m-1}{j}x > 1, \quad \text{or} \quad j > (m+1)\frac{x}{x-1}.$$

He took as an example $m = 1/2$ and $x = 200/199$, in which case the summands do not start getting larger until

$$j > \frac{3}{2}\cdot\frac{200/199}{1/199} = 300.$$

In other words, one must take more than 300 terms of the binomial series expansion of $(1+x)^{1/2}$ before we start to see divergence.

D'Alembert did not do this calculation, but it is easy for us to verify that even though the series expansion of $(1+x)^{1/2}$ at $x = 200/199$ diverges, when we take the first 300 terms, we get a reasonably good approximation: $1.41586\ldots$, as opposed to the true value of $1.41598\ldots$. Using this and other examples, he illustrated how the initial behavior of a series, whether the summands are increasing or decreasing in size, does not indicate whether the series will be convergent or divergent.

The 1768 paper is also interesting because d'Alembert treated the case where $|x| < 1$ in a manner that suggests the confusion that students often exhibit between what is happening to the summands and what is happening to the series. He stated that if $|x| < 1$, then "the series can always be used because it will be convergent at its extremity, and the last term will be infinitely small."[8] While d'Alembert was almost certainly using the word "convergent" to mean that the summands approached zero and not in the modern sense, he appears to have believed that this was sufficient justification to "use" this series to obtain approximations.

Nevertheless, he did come tantalizingly close to a full proof of convergence in the case $|x| < 1$. He used geometric series to put upper and lower bounds on the error when the first $N \geq m + 1$ terms of the binomial series are taken to approximate $(1 + x)^m$ with $-1 < x < 0$.

If $j \geq N \geq m + 1 > 0$ and $-1 < x < 0$, then $\frac{m-j+1}{j}x$, the amount we multiply the previous term to get the summand with x^j, satisfies the inequalities

$$0 < \frac{m - N + 1}{N}x \leq \frac{m - j + 1}{j}x = \frac{j - (m + 1)}{j}|x| \leq |x|.$$

He was able to bound the tail of the series between two geometric series,

$$\left| \sum_{j=N}^{\infty} \frac{m(m-1)\cdots(m-j+1)}{j!}x^j \right| \leq \left| \frac{m(m-1)\cdots(m-N+1)}{N!}x^N \right| \sum_{j=0}^{\infty} |x|^j$$

$$\leq \left| \frac{m(m-1)\cdots(m-N+1)}{N!} \right| \frac{|x|^N}{1 - |x|}$$

$$\left| \sum_{j=N}^{\infty} \frac{m(m-1)\cdots(m-j+1)}{j!}x^j \right| \geq \left| \frac{m(m-1)\cdots(m-N+1)}{N!}x^N \right|$$

$$\sum_{j=0}^{\infty}\left(\frac{m-N+1}{N}x \right)^j$$

$$\geq \left| \frac{m(m-1)\cdots(m-N+1)}{N!} \right| \frac{|x|^N}{1 - \frac{m-N+1}{N}x}.$$

While today we recognize this as a proof of convergence for the case $-1 < x < 0$ because the tail can be brought arbitrarily close to zero by taking N sufficiently large, d'Alembert only saw it as a means of determining bounds for the error.

3.5

Lagrange Remainder Theorem

Joseph Louis Lagrange (1736–1813) was Italian by birth, but spent most of his life in his adopted country, France. He and Pierre-Simon Laplace came to dominate French mathematics in the late eighteenth and early

Figure 3.3. Joseph Louis Lagrange.

nineteenth centuries. Lagrange embraced Euler's approach to calculus. Johann Bernoulli was probably the first to use anything approaching our modern function notation, writing ϕx to denote ϕ as a function of x. But this was not common until Lagrange, who came up with the idea of using parentheses, $\phi(x)$. Lagrange also introduced the use of accent marks to denote the derived functions, $f'(x), f''(x), \ldots$

Lagrange so thoroughly adopted Euler's power series as fundamental to calculus that he defined the derivative in terms of power series. In the very first chapter of his calculus textbook, *Theory of Analytic Functions*, published in 1797,[9] he stated that every function can be expanded in a power series such that

$$f(x+i) = f(x) + pi + qi^2 + ri^3 + \cdots,$$

where $p, q, r, \ldots$ are constants depending on the function f and the value of x. He then defined $f'(x)$ as p. For example, if $f(x) = x^k$, then

$$f(x+i) = (x+i)^k = x^k + kx^{k-1}i + \frac{k(k-1)}{2!}x^{k-2}i^2 + \cdots.$$

Therefore, the derivative of x^k is kx^{k-1}.

Today we use the term *analytic* to describe a function for which such a series exists and converges to the function at all points inside some open interval. Lagrange implicitly assumed that all functions are analytic. We now know that is not true. My favorite example was explained by Cauchy in 1821. It is the function defined to be 0 at $x = 0$ and equal to e^{-1/x^2} at all other values of x.

With a little work and using the limit definition of the derivative, it is possible to show that all derivatives of this function exist and are equal to 0 at $x = 0$. For example, we start with the observation

$$x^2 e^{1/x^2} > x^2 \left(1 + \frac{1}{x^2}\right) > 1.$$

We divide both sides by $|x|e^{1/x^2}$ to get $|x| > \left|x^{-1}e^{-1/x^2}\right|$. We now consider the limit definition of the derivative of e^{-1/x^2} at $x = 0$:

$$0 \leq \left| \lim_{h \to 0} \frac{e^{-1/h^2} - 0}{h} \right| \leq \lim_{h \to 0} |h| = 0.$$

The derivative at $x = 0$ exists and equals 0. Likewise, $f''(0) = f'''(0) = \ldots = 0$ for all higher derivatives.

If this function did have a power series expansion in an open interval containing $x = 0$, then that power series representation would have coefficients that were all 0, which means that this function would be indistinguishable from the constant function $f(x) = 0$. But the fact is that for all values of x other than 0, $e^{-1/x^2} \neq 0$. Despite the fact that all of the derivatives exist at $x = 0$, the power series does not converge to our function at any other points.

Despite the mistaken assumption that all functions are analytic, *Theory of Analytic Functions* made many important advances in our understanding of calculus. One result that first appeared in this book is the Lagrange remainder theorem. By providing explicit bounds for the difference between a Taylor polynomial and the function it approximates, this marked an important step in our understanding of issues of convergence.

Lagrange Remainder Theorem. *If f is analytic in an open interval that contains a, then for any x in this interval and any $n \geq 0$ the difference between $f(x)$ and the nth Taylor polynomial for f about a can be expressed in terms of the $n + 1$st derivative of f at some point, say c, between a and x,*

$$f(x) - \left(f(a) + f'(a)(x-a) + \frac{f''(a)}{2!}(x-a)^2 + \cdots + \frac{f^{(n)}(a)}{n!}(x-a)^n \right)$$

$$= \frac{f^{(n+1)}(c)}{(n+1)!}(x-a)^{n+1}.$$

Note that the case $n = 0$ with $x = b$ is the mean value theorem,

(3.11) $f(b) - f(a) = f'(c)(b-a)$ or $\dfrac{f(b) - f(a)}{b-a} = f'(c).$

In fact, the Lagrange remainder theorem is nothing more than an extended mean value theorem.

Lagrange's proof follows directly from the intuitively obvious result, sometimes referred to as the *increasing function theorem*, that if $f'(x) \geq 0$ on the interval $[a, b]$, then f is increasing on this interval and, in particular, $f(b) \geq f(a)$. In 1967, Lipman Bers published a short piece in the *American Mathematical Monthly*[10] observing that the mean value theorem follows easily from the increasing function theorem. This is ironic because almost all calculus textbooks derive the increasing function theorem from the mean value theorem, and in many cases it is the increasing function theorem that is the more useful of the two. Bers noted that his derivation of the mean value theorem from the increasing function theorem was "hardly new," and he asked for a reference, not realizing that this was the approach used by Lagrange.

Lagrange used the increasing function theorem as follows. Let M be the maximum value of f' on $[a, b]$ and let $N(x) = Mx - f(x)$. Then $N'(x) = M - f'(x) \geq 0$. Therefore $N(b) \geq N(a)$, or

$$Mb - f(b) \geq Ma - f(a), \quad \text{or} \quad \frac{f(b) - f(a)}{b - a} \leq M.$$

Similarly, if m is the minimum value, then $f'(x) - m \geq 0$ and

$$f(b) - mb \geq f(a) - ma \quad \text{or} \quad \frac{f(b) - f(a)}{b - a} \geq m.$$

This is where Lagrange stopped, because he had succeeded in putting bounds on the error term. Today, knowing that any derivative must satisfy the intermediate value property, we can conclude that there is some c in the interval for which the average rate of change equals the instantaneous rate of change.

The general case of Lagrange's remainder theorem is not much harder. Let g_n denote the difference between f and the nth Taylor polynomial approximation to f,

$$g_n(x) = f(x) - \left(f(a) + f'(a)(x - a) + \frac{f''(a)}{2!}(x - a)^2 + \cdots \right.$$

$$\left. + \frac{f^{(n)}(a)}{n!}(x - a)^n \right).$$

The nth derivative of $g_n(x)$ is $f^{(n)}(x) - f^{(n)}(a)$, and the $n + 1$st derivative is $f^{(n+1)}(x)$. Let M be an upper bound for $f^{(n+1)}(x)$ so that $M - f^{(n+1)}(x) = M - g_n^{(n+1)}(x)$ is always positive. By the increasing function theorem,

$$Mx - g_n^{(n)}(x) \geq Ma - g^{(n)}(a) = Ma, \quad \text{or} \quad M(x - a) - g_n^{(n)}(x) \geq 0.$$

Because $M(x - a) - g_n^{(n)}(x)$ equals 0 at $x = a$, we also have

$$\frac{M}{2}(x - a)^2 - g_n^{(n-1)}(x) \geq 0.$$

But this function is also 0 at $x = a$. Repeating this process through the anti-derivatives of $g_n^{(n)}$, we eventually reach

$$\frac{M}{(n + 1)!}(x - a)^{n+1} - g_n(x) \geq 0,$$

showing that $g_n(x)$ is bounded above by $M(x - a)^{n+1}/(n + 1)!$. A similar argument works for the lower bound. Therefore,

$$m\frac{(x-a)^{n+1}}{(n+1)!} \leq g_n(x) \leq M\frac{(x-a)^{n+1}}{(n+1)!},$$

where m and M are, respectively, the lower and upper bounds on $f^{(n+1)}(x)$. Therefore, there is a value c between a and b for which

$$g_n(x) = f^{(n+1)}(c)\frac{(x-a)^{n+1}}{(n+1)!},$$

the result that we sought to prove.

Remember that all of this rests on the increasing function theorem. Lagrange realized that he had to prove this. His proof was very insightful for its time but would not pass muster today.

Because $f'(a)$ is positive, there must be an interval of some length, say i, to the right of a where $f(x)$ is strictly larger than $f(a)$. Lagrange justified this by appeal to the Taylor series expansion of f. Today we would justify it by the limit definition of the derivative. That is fine. But he then asserted that

$$f(a) < f(a+i) < f(a+2i) < f(a+3i) < \cdots < f(a+ni),$$

which eventually must include $f(b)$. The problem is that he assumed that the same length, i, that works for a will work across the entire interval from a to b. This was common. A quarter of a century later, Cauchy would make the same assumption when he attempted to prove the mean value theorem. It raises a tricky point because if f is differentiable with a continuous derivative on the closed interval $[a, b]$, then there is indeed such a minimal interval length, i. But this is a subtle and difficult result that would not be fully understood for almost a century.

The increasing function theorem is not difficult to prove, and Lipman Bers provided a proof which I reproduce here. But this is a proof that would never have occurred to Lagrange or Cauchy. It is unquestionably a product of the late nineteenth century.

Assume that $f'(x) > 0$ for all x between a and b, but there is a value p in (a, b) for which there is at least one x between a and p with $f(x) \geq f(p)$. Let S be the set of all such values of x, $a < x < p$, and let q be the smallest value in $[a, p]$ that is greater than or equal to every element of S. Because $f'(p) > 0$, q cannot equal p. If $f(q) \geq f(p)$, then since

$f'(q) > 0$, there are points of S to the right of q, a contradiction. If $f(q) < f(p)$, then q is not in S and, by continuity, there are no points of S near and to the left of q, another contradiction.

I must conclude with a nod to the proof of the mean value theorem that is now ubiquitous. One begins by subtracting the linear function that passes through $(a, f(a))$ and $(b, f(b))$,

$$g(x) = f(x) - \frac{f(b) - f(a)}{b - a}(x - a).$$

The resulting function g must have a local extremum between a and b, and at that extremum the derivative is zero, $g'(c) = 0$. Because this derivative equals the derivative of the original function minus the slope of the linear function, which is also the average rate of change of f, the two are equal at that point,

$$0 = g'(c) = f'(c) - \frac{f(b) - f(a)}{b - a} \implies f'(c) = \frac{f(b) - f(a)}{b - a}.$$

To the best of my knowledge, this was first discovered by Ossian Bonnet (1819–1892) and first published by Joseph Alfred Serret in his calculus textbook of 1868. It is brilliantly simple, but the fact that it took almost 70 years before it was widely known suggests that it is not an intuitively obvious approach.

3.6
Fourier's Series

As we saw in the last sections of chapter 2, partial differential equations play an essential role in understanding and exploring our physical world. The derivation we saw in section 2.10 with the vibrating string is a common approach to solving a partial differential equation involving functions of several variables. We simplify the problem to the case where the solution is a product of functions of fewer variables and then use a sum of these products to build the full solution. The coefficients for this sum are determined by taking the function that describes the boundary condition and expanding it in terms of an appropriately chosen family of functions. For

Figure 3.4. Watercolor caricature of Joseph Fourier by Julien-Léopold Boilly, 1820.

the vibrating string, the set of functions that can be used to describe the initial condition is

$$\{\sin(m\pi x) \mid m \geq 1\}.$$

We can use any finite linear combination of these functions as an initial condition, but there was often heated debate in the eighteenth century whether one could employ infinitely many of them.

In the early nineteenth century, Joseph Fourier (1768–1830) sought to explain the propagation of heat in a solid body. If one applies heat to one end of a long, thin metal plate, is it possible to predict the amount of heat at different points in the plate? His solution to this problem was roundly rejected by the leading mathematicians of the time because it challenged much of what they thought they knew about infinite series.

Fourier had a varied and interesting career. The twelfth of fifteen children of a tailor in Auxerre, France, he had begun training for the priesthood but also held an interest in mathematics. When, in 1794, the First Republic established the École Normale "to which will be called from all parts of the Republic, in order to learn from the most skillful professors in all fields, the art of teaching,"[11] he went to Paris as one of its pupils. There he came in contact with the most famous Frenchmen working in mathematics at that time, including Lagrange, Laplace, and Monge. By the fall of 1795, he was teaching at the École Polytechnique, the elite technical school established by the First Republic to provide an engineering education to prospective officers.[12]

In 1798, Fourier was chosen as a member of the corps of scientists who accompanied Napoleon during his invasion of Egypt. There he helped to establish the Egyptian Scientific Institute and engaged in archaeological explorations, which would lead in later years to his role in the production of *Description of Egypt*, a comprehensive catalog of both ancient and modern Egypt. After returning to France in 1801, he was appointed prefect (governor) of the Department of Isère in the French Alps, where he was tasked with building an improved highway from Grenoble over the Alps to Turin in Italy.

Fourier continued to work in science and mathematics. In 1807 he presented his work, *On the propagation of heat in solid bodies*, to the Institut de France. This is where he unleashed on the mathematical community what today we know as Fourier series, creating a firestorm of controversy.

The starting point for Fourier was the recognition that heat travels through metal like an incompressible fluid. This means that in a steady state, the amount of heat H at a given point (x, y) must satisfy the divergence equation

$$(3.12) \qquad \frac{\partial^2 H}{\partial x^2} + \frac{\partial^2 H}{\partial y^2} = 0,$$

the two-dimensional version of Laplace's equation (2.14). The initial condition describes how the heat is applied along the edge where $y = 0$ and x lies between 0 and π. Perhaps surprisingly, the set of solutions in x once again employs sine functions, $\{\sin(mx) \mid m \geq 1\}$.

Fourier now attempted to find the combination of these sine functions that would yield a uniform value of 1 for x between 0 and π. Using the trigonometric identity

$$\sin(kx)\ \sin(mx) = \frac{1}{2}\left(\cos\left((k - m)x\right) - \cos\left((k + m)x\right)\right),$$

Fourier observed that when k and m are integers,

$$(3.13) \quad \int_0^\pi \sin(kx)\ \sin(mx)\ dx = \frac{1}{2}\int_0^\pi \cos\left((k - m)x\right) - \cos\left((k + m)x\right)\ dx$$

$$= \begin{cases} \pi/2, & \text{if } k = m, \\ 0, & \text{if } k \neq m. \end{cases}$$

He now assumed that the function that is identically equal to 1 for x between 0 and π can be expressed as a sum of these sine functions,

$$1 = a_1 \sin x + a_2 \sin 2x + a_3 \sin 3x + \cdots = \sum_{m=1}^{\infty} a_m \sin mx.$$

He used equation (3.13) to find the coefficients:

$$\int_0^\pi \sin kx \cdot 1 \, dx = \int_0^\pi \sin kx \left(\sum_{m=1}^{\infty} a_m \sin mx \right) dx$$

$$= \sum_{m=1}^{\infty} a_m \int_0^\pi \sin kx \sin mx \, dx$$

$$= \frac{\pi}{2} a_k.$$

Because

$$\int_0^\pi \sin kx \, dx = \frac{1}{k} (-\cos kx) \Big|_0^\pi$$

$$= \frac{1}{k} (1 - \cos k\pi)$$

$$= \begin{cases} 2/k, & \text{if } k \text{ is odd,} \\ 0, & \text{if } k \text{ is even,} \end{cases}$$

it follows that $a_k = 4/k\pi$ when k is odd; $a_k = 0$ when k is even. This finally yields

(3.14) $$1 = \frac{4}{\pi} \left(\sin x + \frac{1}{3} \sin 3x + \frac{1}{5} \sin 5x + \cdots \right), \qquad \text{for } 0 < x < \pi.$$

Note that when $x = \pi/2$, this becomes

$$\frac{\pi}{4} = 1 - \frac{1}{3} + \frac{1}{5} - \frac{1}{7} + \cdots,$$

the series we saw in equation (2.8). If we plot the partial sums obtained from the first three, six, fifteen, or fifty summands of the series in equation (3.14)

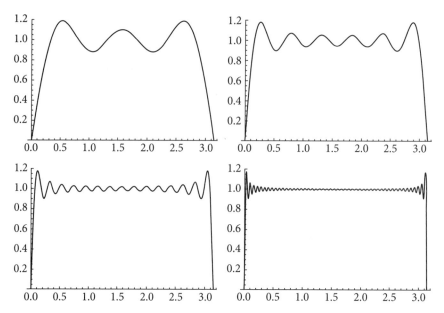

Figure 3.5. Graphs of the first three, six, fifteen, and fifty terms of the Fourier series on the right side of equation (3.14).

(Figure 3.5), we see that this series does seem to be approaching the constant function 1.

But there are serious problems with equation (3.14). Note that it is patently false when $x = 0$ or π. If we try to extend this series to the interval between π and 2π by adding π to x, we see that $\sin k(x + \pi) = -\sin kx$ when k is odd, so this series would have to equal -1 between π and 2π (see Figure 3.6). If equation (3.14) is valid, we really do need to restrict x to the interval between 0 and π.

There is more wrong with this infinite sum of sine functions. The sine is a continuous function and, as everyone knows, the sum of continuous functions is again continuous. This function, which jumps between $+1$, 0, and -1, is clearly not continuous (see Figure 3.7). It is also not differentiable. If we try differentiating it, we get

$$\frac{4}{\pi}(\cos x + \cos 3x + \cos 5x + \cdots),$$

which does not even converge at $x = \pi/4$. If the Fourier series were equal to 1 between 0 and π, its derivative would be 0 at $x = \pi/4$.

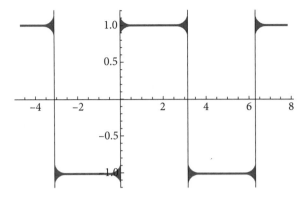

Figure 3.6. Graph of the first fifty terms of the Fourier series on the right side of equation (3.14) over the domain $-3\pi/2 \leq x \leq 5\pi/2$.

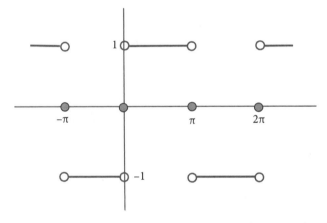

Figure 3.7. Graph of the Fourier series on the right side of equation (3.14) over the domain $-3\pi/2 \leq x \leq 5\pi/2$.

It should come as no surprise that the greatest French mathematicians of the time, including Pierre-Simon Laplace and Joseph-Louis Lagrange, rejected this work. Nonetheless, equation (3.14) is, in fact, correct.

As mathematicians came to realize in the succeeding decades, an infinite sum of continuous functions is not necessarily continuous. An infinite sum of differentiable functions is not necessarily differentiable, and, even if it is, its derivative is not necessarily equal to the sum of the derivatives of the individual summands. Settling the doubts and uncertainties created by Fourier's series would become one of the great projects consuming the nineteenth century. As the mathematicians of this era focused on what could go wrong, they found continuous functions that were nowhere differentiable, derivatives that could not be integrated, and integrals that could not be differentiated, each new discovery forcing them to re-examine their basic assumptions. By the end of this process, calculus would be completely transformed, resting on clear and totally unambiguous definitions, built up painstakingly through lemmas, propositions, and theorems of increasing sophistication. By the late nineteenth century, so much attention was focused on what could go wrong that Henri Poincaré, one of the greatest mathematicians of the time, wrote with despair,

> In earlier times, when we invented a new function it was for the purpose of some practical goal. Today, we invent them expressly to show the flaws in our forefathers' reasoning, and we draw from them nothing more than that. (Poincaré, 1889)

With the final defeat of Napoleon in 1815, Fourier returned to Paris. In 1822, he was elected secretary of the Académie des Sciences. Shortly afterwards he published his definitive work on heat flow and Fourier series, *Théorie analytique de la chaleur* (Analytic theory of heat). In the remaining years before his death in 1830, he mentored Sophie Germain[13] and many of the young scientists who would leave their mark on nineteenth-century mathematics: Claude Navier, Charles Sturm (1803–1855), Gustav Dirichlet (1805–1859), and Joseph Liouville (1809–1882). In addition to encouraging them in their pursuit of mathematics, he regaled them with tales of his adventures in Egypt.

Chapter 4

THE ALGEBRA OF INEQUALITIES

In 1825, the Norwegian Niels Henrik Abel (1802–1829) visited the mathematical community in Berlin and arrived at a most astounding conclusion. In a letter sent to his teacher Bernt Michael Holmboe in Christiania,[1] he wrote,

> My eyes have been opened in the most surprising manner. If you disregard the very simplest cases, there is in all of mathematics not a single infinite series whose sum has been rigorously determined. In other words, the most important parts of mathematics stand without foundation. It is true that most of it is valid, but that is very surprising. I struggle to find a reason for it, an exceedingly interesting problem. (Ore 1974, p. 97; see also Stubhaug 1996, p. 343)

Abel was overly pessimistic about the status of convergence of infinite series. As we have seen in section 3.4, d'Alembert had made the first steps toward the ratio test. The error bound established by Lagrange is often an effective tool for proving the convergence of Taylor series. In 1812, Carl Friedrich Gauss published an extensive treatise on a type of Taylor series known as *hypergeometric series*,[2] finding not just the radius of convergence but laying out necessary and sufficient conditions for convergence at the endpoints.

Nevertheless, there was great confusion within the mathematical community of the time over whether convergence was necessary and how one might determine whether or not a series converged. By 1825, Fourier series had been accepted as legitimate, but no one was capable of proving that the series in equation (3.14) converged.

As Abel would discover when he traveled on to Paris, one man, Augustin-Louis Cauchy, was making significant progress toward the solution of this problem. Cauchy employed the algebra of inequalities to establish the criteria for the convergence of infinite series and, for the first time, provide definitions of the derivative and integral that would enable the construction of proofs of their properties.

This chapter will trace the history of the limit concept, leading up to Cauchy's insights. A consequence of the light he was able to shed on the fundamentals of calculus would reveal unexpected questions, some of which we will address in this chapter, many of which we will postpone until chapter 5 on analysis.

4.1
Limits and Inequalities

Limits have lurked in the background of all we have studied so far. Beginning with Archimedes and continuing throughout our story, we have seen the conflict between an intuitive use of infinitesimals or indivisibles and infinities on the one hand and a rigorous treatment without resort to such "fantasies" on the other. One problem with Euclidean rigor was that the arguments had to be tailor-made. One showed the validity of a particular answer by demonstrating that the true value could be neither larger nor smaller.

Inspired by Kepler and accelerating through the work of Torricelli, Wallis, the Bernoullis, and Euler, arguments based on infinitesimals became ever more common. They were almost always accompanied by unease over their use. This was frequently assuaged by pointing out that the successive approximations arising from progressively smaller increments could be brought as close as one might wish to the desired value.

This is implicit in Beeckman's argument (section 1.8) that distance traveled is the area under the line that records velocity. We see it in Wallis. He argued that when the ratio

$$\frac{0^2 + 1^2 + 3^2 + \cdots + \ell^2}{\ell^2 + \ell^2 + \ell^2 + \cdots + \ell^2} = \frac{1}{3} + \frac{1}{6\ell}$$

is "continued to infinity," the "excess over one-third is continually decreased, in such a way that at length it becomes less than any assignable quantity."[3]

Newton began Book I of the *Principia* with a lemma that defines what he meant by a limit:

> Quantities, and also ratios of quantities, which in any finite time constantly tend to equality, and which before the end of that time approach so close to one another that their difference is less than any given quantity, become ultimately equal. (Newton, 1687, p. 433)

His "proof" illustrates the connection to Euclidean rigor:

> If you deny this, let them be ultimately unequal, and let their ultimate difference be D. Then they cannot approach so close to equality that their difference is less than the given difference D, contrary to the hypothesis.

Sometime before 1765, d'Alembert wrote the entry on *Limit (Mathematics)* for the *Encyclopédie*. It begins with the following definition:

> One says that a magnitude is the *limit* of another magnitude when the second may approach the first more closely than by a given quantity, as small as one wishes, moreover without the magnitude which approaches being allowed ever to surpass the magnitude that it approaches; so that the difference between such a quantity and its *limit* is absolutely unassignable. (As translated in Stedall, 2008, p. 297)

Grabiner has noted the curious restriction that the approaching variable is only allowed to come from one direction and observed that this was common throughout the eighteenth century. It was so ingrained that when, in 1795, Lhuilier discussed the convergence of an alternating series, he needed a special definition of limit to handle this case.[4]

To those who are conversant with the modern epsilon-delta definition of limit, all of this looks pretty close to the way we understand limits today. It is couched in the language of approaching a value. The critical piece— that we can force the difference between the approximating values and the limit to be less than any given quantity—looks like the statement that for any $\epsilon > 0$, we can get this difference less than ϵ. In fact, none of this appears to be very different from Cauchy's definition in *Cours d'analyse* of 1821:

When the values successively attributed to the same variable approach indefinitely to a fixed value, in such a way as to end by differing from it as little as one wishes, this last is called the *limit* of all the others. (As translated in Stedall, 2008, p. 300)

As Grabiner has pointed out, two things distinguish Cauchy's treatment of limits from what came before: the translation of this verbal description into a practical algebra of inequalities and the ability to then use these inequalities to prove basic results about calculus.

4.2
Cauchy and the Language of ϵ and δ

As with anyone leading the way into unfamiliar territory, Cauchy's accomplishments were insightful, yet frequently incomplete and misleading. After Abel encountered Cauchy's work on calculus, he wrote back to Holmboe,

Cauchy is crazy, and there is no way of getting along with him, even though right now he is the only one who knows how mathematics should be done. What he is doing is excellent, but very confusing.

Cauchy was born in Paris, a month and a week after the storming of the Bastille. The family had to flee Paris because of his father's high position in the city police, but they returned after the execution of Robespierre in 1794. Growing up, both Laplace and Lagrange were family friends who encouraged Cauchy's mathematical talent. In 1805 he entered the École Polytechnique where he earned an engineering degree and a commission as lieutenant in Napoleon's army.

In 1810 he was sent to Cherbourg on the English channel to work on the upgrading of its harbor in preparation for launching the fleet that would invade England.[5] In September 1812, exhausted from overwork, he returned to Paris. Less than two months later, he submitted the 84-page manuscript that in many ways marks the beginning of linear algebra, "Memoir on functions whose values are equal but of opposite sign when two of their variables are interchanged." Among many important results in this paper, Cauchy gave the first proof that the determinant of a product of two square matrices is equal to the product of the determinants.[6]

Figure 4.1. Augustin-Louis Cauchy.

Cauchy was hired to teach at the École Polytechnique in 1815. In 1821, he published the first volume of a calculus textbook for his students, *Cours d'Analyse* (Course of analysis),[7] marking the beginnings of a precise and rigorous treatment of calculus. He would further develop these ideas in *Résumé des leçons ... sur le Calcul Infinitésimal* (Summary of lectures on infinitesimal calculus).[8] In the introduction to *Cours d' Analyse*, he expressed his dissatisfaction with the freewheeling style employed by Euler and Lagrange,

> As for my methods, I have sought to give them all of the rigor that one demands in geometry in such a way as never to rely on explanations drawn from algebraic technique. Such reasons cannot be considered, in my opinion, except as heuristics that will sometimes suggest the truth, but which accord little with the accuracy that is so praised in the mathematical sciences.

Cauchy began *Cours d'Analyse* with two examples showing how limits are to be considered,

$$\lim_{x\to 0} \frac{\sin x}{x} \quad \text{and} \quad \lim_{x\to 0} (1+x)^{1/x}.$$

While epsilon and delta do not make their formal appearance until the 1823 *Calcul Infinitésimal*, the algebra of inequalities is very apparent. For the first

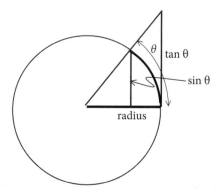

Figure 4.2. Circle representation of trigonometric functions showing that when radius and arc length are measured in the same units, $\sin\theta \leq \theta \leq \tan\theta$.

limit, he observed that for small values of x,

$$\sin x < x < \tan x,$$

a fact that is obvious from Figure 4.2. Therefore

(4.1) $$1 = \frac{\sin x}{\sin x} > \frac{\sin x}{x} > \frac{\sin x}{\tan x} = \cos x.$$

Clearly, the value that we assign to the limit of $(\sin x)/x$ cannot be larger than 1. But it also cannot be less because $\cos x$ can be brought as close to 1 as we want by taking x sufficiently close to 0.

In the language of ϵ and δ, the limit,

$$L = \lim_{x \to a} f(x),$$

exists if we can force $f(x)$ to be as close to L as we wish by controlling the distance between x and a:

> Given any $\epsilon > 0$, we can force $|L - f(x)| < \epsilon$ by finding a $\delta > 0$ so that L and $f(x)$ are this close when $0 < |x - a| < \delta$.

The limit as x approaches 0 of $(\sin x)/x$ is 1 because given any $\epsilon > 0$, we can force $\cos x$ to be within ϵ of 1 by taking x sufficiently close—not equal—to 0.[9] From inequality (4.1), the distance between $(\sin x))/x$ and 1 is less than the distance between $\cos x$ and 1.

Cauchy's second example, $\lim_{x \to 0}(1+x)^{1/x}$, is trickier. He first replaced x by $1/m$, where m is a positive integer that can be as large as we wish. Then he used the binomial expansion,

$$\left(1 + \frac{1}{m}\right)^m = 1 + m\frac{1}{m} + \frac{m(m-1)}{2!}\left(\frac{1}{m}\right)^2$$

$$+ \frac{m(m-1)(m-2)}{3!}\left(\frac{1}{m}\right)^3 + \cdots + \left(\frac{1}{m}\right)^m$$

$$= 1 + 1 + \frac{1}{2!}\left(1 - \frac{1}{m}\right) + \frac{1}{3!}\left(1 - \frac{1}{m}\right)\left(1 - \frac{2}{m}\right) + \cdots + \frac{1}{m^m}.$$

Because each summand is positive, this is clearly larger than 2 when $m \geq 2$. Recognizing that

$$\frac{1}{k!} \leq \frac{1}{2^{k-1}}$$

for $k \geq 1$, he found an upper bound of 3,

$$2 \leq \left(1 + \frac{1}{m}\right)^m < 1 + 1 + \frac{1}{2} + \frac{1}{2^2} + \frac{1}{2^3} + \cdots = 3.$$

As m gets larger, each summand after the second gets larger, and we have more summands, so the value of $(1 + 1/m)^m$ is increasing. But it is also bounded above by 3. There must be some value L, located between 2 and 3, to which this sequence converges. This value is the constant that Euler dubbed e. The assertion that a bounded increasing sequence must converge is actually a subtle point to which mathematicians would return in the second half of the nineteenth century and to which we will return in section 4.3. It is worth recognizing that to Cauchy and his contemporaries, this assertion seemed too obvious to require careful attention. Also note that there is no way to use ϵ or δ here, because there is no explicit value of L proposed by Cauchy (to call it e gives it a name, not a value).

We do need to cut Cauchy some slack. As the first person to try to build this foundation for calculus with the full rigor of Euclidean geometry, Cauchy left inevitable gaps. The nature of this gap is illustrated in the following definition of the limit of a function, which is equivalent to the ϵ-δ definition, but without the epsilons and deltas. We consider all punctured open intervals around a (an interval of the form $(c, a) \cup (a, d)$, $c < a < d$) intersected with the set of values on which f is defined. For each punctured open interval around a, we note the greatest and least values of f.[10] If there is a unique number less than or equal to all of these greatest values and greater than or equal to all least values, then the limit exists and is given by that number.

These definitions of limit provide us with our modern interpretation of instantaneous velocity, equivalent to the derivative of the position function.

Definition 4.1 (Derivative or Instantaneous Velocity, I). *In the language of epsilons and deltas, if $s(t)$ is the position at time t, then the derivative of s at $t = a$, which is the instantaneous velocity v at time a, is defined to be the unique value, $v(a)$, such that for any $\epsilon > 0$ there is a response $\delta > 0$ for which $0 < |t - a| < \delta$ implies that*

$$\left| \frac{s(t) - s(a)}{t - a} - v(a) \right| < \epsilon.$$

Definition 4.2 (Derivative or Instantaneous Velocity, II). *In plain English, for each open interval I containing a, we consider the average rates of change or average velocities between a and the other points in the interval. Let M_I be the maximum of these average velocities and m_I the minimum. As the intervals get shorter, the distance between M_I and m_I should decrease. The derivative or instantaneous velocity at time $t = a$ is the unique value that is less than or equal to M_I and greater than or equal to m_I for every open interval I.*

While they are equivalent, there is an important distinction between these two definitions of the derivative. In the first, we began with the assumption that there is a candidate for the limit and described the conditions it must satisfy. In the second, we described a set of numbers: those that are less than or equal to M_I and greater than or equal to m_I for every open interval I containing a. If and only if this set contains exactly one element, then the limit exists and equals that element. This leads us to two important questions that will open up the world of nineteenth-century analysis:

(1) When does such an element exist?
(2) When is it unique?

In Cauchy's second example, we know that the greatest value is less than or equal to 3. But for the least values, all we know is that they are larger than 2 and increasing as m increases. Neither existence nor uniqueness is obvious.

4.3
Completeness

Cauchy appeared unaware of the difficulty of justifying existence of the limit, simply asserting that for an infinite sequence, if for every $\epsilon > 0$, there is a response M such that all of the terms at or beyond the Mth lie within ϵ of each other, then the sequence converges. Today we call such a sequence a *Cauchy sequence*. The sequence $(1 + 1/m)^m$, $m \geq 1$ is Cauchy because of two properties we have already established. First, since all terms from the third on lie strictly between 2 and 3, there should be some least number that is greater than or equal to every term of the sequence, called the *least upper bound*. This means that for any $\epsilon > 0$, there must be some term of the sequence that lies within ϵ of this upper bound (otherwise, we could pick a smaller upper bound). Second, because the sequence is increasing, once one term is within ϵ of this bound, all succeeding terms are also within ϵ. This guarantees the uniqueness of the limit, but what about existence? How do we know that there actually is a least upper bound?

To understand what is needed, it is useful to recognize that a Cauchy sequence does not always converge if we restrict our attention to rational numbers. The square root of 2 cannot be written as a ratio of two integers, p/q. But we can find a sequence of rational numbers that does converge to $\sqrt{2}$. The classic example, well-known to Archimedes, is

$$\frac{1}{1}, \frac{3}{2}, \frac{7}{5}, \frac{17}{12}, \frac{41}{29}, \frac{99}{70}, \ldots,$$

where a/b is succeeded by $(a + 2b)/(a + b)$.[11] This is a Cauchy sequence that lies entirely within the set of rational numbers. But if we can only use rational numbers, then it has no limit.

Until well into the nineteenth century, those working with sequences and series simply assumed that every bounded increasing sequence converges and every Cauchy sequence converges. It is a straightforward exercise to show that these two qualities are equivalent.[12] Today we refer to this property of the real numbers as *completeness*; the set of real numbers is *complete*.

Almost all of the tests of convergence for infinite series rely on demonstrating that the series in question is Cauchy. These include the ratio, root, and integral tests, which prove convergence by comparing the series in question to a series that is known to converge. Convergence of a series means convergence of the sequence of partial sums, $S_n = \sum_{j=1}^{n} a_j$. A series is Cauchy if for each $\epsilon > 0$ there is a response N so that any two partial sums beyond the Nth differ by less than ϵ. This means that if $n > m \geq N$, then

$$|S_n - S_m| = \left| \sum_{j=1}^{n} a_j - \sum_{j=1}^{m} a_j \right| = |a_{m+1} + a_{m+2} + \cdots + a_n| < \epsilon.$$

The comparison test states that if $\sum a_j$ and $\sum b_j$ are both sums of positive terms, $0 \leq a_j \leq b_j$ for all j, and if $\sum b_j$ is known to converge, then $\sum a_j$ must converge. The proof relies on the convergence of Cauchy sequences. As Cauchy proved, if an infinite series converges, then its partial sums form a Cauchy sequence. Given any $\epsilon > 0$, there is an M such that $k > j \geq M$ implies that $0 \leq b_{j+1} + b_{j+2} + \cdots + b_k < \epsilon$. Since $0 \leq a_{j+1} + a_{j+2} + \cdots + a_k \leq b_{j+1} + b_{j+2} + \cdots + b_k$, the partial sums of $\sum a_j$ are also Cauchy, and thus also convergent. Note that we could only remove the absolute value signs because everything is positive.

In the second half of the nineteenth century, several mathematicians struggled with the question of how to define the real numbers to guarantee completeness. Richard Dedekind (1831–1916) was the first to come up with a construction. He presented it in a course at the Zürich Polytechnique (today ETH Zürich) in 1858, although he did not publish it until 1872. The idea is to represent each real number as a *cut* on the real number line, defined by two sets of rational numbers, A and B, such that each point of A is strictly less than each point of B and for any positive distance $\epsilon > 0$, one can find an $a \in A$ and a $b \in B$ such that $|a - b| < \epsilon$. Today, such a pair of sets is known as a *Dedekind cut*.

Several mathematicians, among them Charles Méray (1835–1911), Eduard Heine (1821–1881), and Georg Cantor (1845-1918), came up with the idea that amounts to defining each real number as an equivalence class of Cauchy sequences of rational numbers. Given two Cauchy sequences of rational numbers, $(s_n)_{n=1}^\infty$ and $(t_n)_{n=1}^\infty$, they are equivalent if the sequence of differences converges to 0:

$$\lim_{n \to \infty} (s_n - t_n) = 0.$$

One can simplify this and only consider Cauchy sequences formed by the partial sums of the form

$$s_n = a_0 + \sum_{j=1}^{n} \frac{a_j}{10^j},$$

where a_0 is an integer and each a_j, $j > 0$, is an integer from the set $\{0, 1, 2, \ldots, 9\}$. The partial sums of such a series are necessarily rational, and this is a Cauchy sequence. What we have done is to identify each real number with the equivalence class of its decimal representations. Of course, the only time a real number has two distinct representations is when one of them ends in an infinite sequence of repeating 9's. Thus, this definition of real numbers forces the equality of $0.9999\ldots$ and 1.

4.4
Continuity

Armed with his definition of limit, Cauchy could now define the derivative in a precise manner. But first he dealt with continuity. Continuity is a curious concept within calculus because no one worried about it until the early nineteenth century, when suddenly it took on an enormous importance. If all functions can be expressed as power series, continuity is not an issue. As scientists began to expand their understanding of what a function could be, issues of continuity took on increased importance, especially the intermediate value property that asserts that over an interval $[a, b]$ a continuous function, f, hits every value between $f(a)$ and $f(b)$. Lagrange wrestled with a proof that all polynomials are continuous for just this reason. It arose in one of the most important problems of the late eighteenth century, the Fundamental Theorem of Algebra.

This theorem, whose first generally accepted proof was given by Carl Friedrich Gauss, states that every polynomial with real or complex coefficients will always have a real or complex root. Once we know that there is at least one root, it is straightforward to deduce that a polynomial of degree $n \geq 1$ will have exactly n real or complex roots (counted by multiplicity), because if $p(x)$ is a polynomial of degree n with root r, then $p(x)/(x - r)$ is a polynomial of degree $n - 1$. To establish the existence of such a root, we need to know that a real-valued function that takes on both positive and negative values must equal 0 somewhere. Cauchy devoted an entire chapter of the *Cours d' Analyse* to his own take on this proof, relying on this intermediate value property of continuous functions.

Cauchy defined the function f to be continuous at a if "$f(x + a) - f(x)$ decreases indefinitely with a."[13] In other words, $\lim_{a \to 0} f(x + a) - f(x) = 0$. He then considered the case of a continuous function that takes on both positive and negative values: $f(x_0) > 0, f(X_0) < 0$, where $x_0 < X_0$.

Theorem 4.1 (Intermediate Value Theorem). *If f is a continuous function that takes on both positive and negative values, say $f(x_0) > 0$, $f(X_0) < 0$, where $x_0 < X_0$, then there is at least one value of x between x_0 and X_0, say $x = a$, such that $f(a) = 0$.*

Proof. We subdivide the interval $[x_0, X_0]$ into m smaller intervals, each of length $(X_0 - x_0)/m$. If f is not zero at any of the endpoints of the new subintervals, then for at least one of these subintervals, f is positive at the left-hand endpoint and negative at the right-hand endpoint. Call this new interval $[x_1, X_1]$. We now subdivide this and find an interval only $1/m^2$ times the size of the original interval, say $[x_2, X_2]$, on which the function is positive at x_2 and negative at X_2, or f is zero at one of the endpoints. If we never encounter an endpoint where f is zero, then we continue this process indefinitely, creating two sequences, $x_0 \leq x_1 \leq x_2 \leq \cdots$ and $X_0 \geq X_1 \geq X_2 \geq \cdots$ that come arbitrarily close to each other. At each stage, $x_k < X_k$ and $X_k - x_k = (X_0 - x_0)/m^k$. These sequences, $x_0 \leq x_1 \leq x_2 \leq \cdots$ and $\cdots \leq X_2 \leq X_1 \leq X_0$, must converge to a single value that Cauchy denoted as a. By the continuity of f,

$$\lim_{k \to \infty} f(x_k) = f(a) = \lim_{k \to \infty} f(X_k).$$

Because $f(x_k) \geq 0$ for all k, the first limit implies that $f(a) \geq 0$. Because $f(X_k) \leq 0$ for all k, the second limit implies that $f(a) \leq 0$. Together these imply that $f(a) = 0$.   □

Cauchy then pointed out that a continuous function f must take on any intermediate value, b, because $f(x) - b$ takes on both positive and negative values. Where $f(x) - b$ is zero, $f(x)$ equals b.

As important as the intermediate value theorem was to Cauchy, continuity took on an even more important role when he recognized that he could establish the existence of the definite integral for *any* continuous function, regardless of whether it possessed a power series representation.

Before we tackle Cauchy's proof that every continuous function is integrable, we need to explore the notion of continuity. Note that our definition of continuity is expressed in terms of what happens at a single value of x: f is continuous at $x = a$ if $\lim_{x \to a} f(x)$ and $f(a)$ both exist and are equal. This creates some apparent paradoxes. The first is a function that is continuous at $x = 0$ and nowhere else:

$$f(x) = \begin{cases} x, & \text{if } x \text{ is rational,} \\ 0, & \text{if } x \text{ is irrational.} \end{cases}$$

It is continuous at 0 because $\left| f(x) - f(0) \right|$ is either 0 or $|x|$. Given any $\epsilon > 0$, for $|x - 0| < \epsilon$ we also have $\left| f(x) - f(0) \right| \leq \epsilon$. At any other value of x, choose an $\epsilon < |x|$. No matter how small an interval around x we take, it will contain both rational and irrational numbers. If x is rational, choose x_1 to be irrational so that $|f(x) - f(x_1)| = |x| > \epsilon$. If x is irrational, choose x_1 to be rational and slightly larger than x in magnitude so that $|f(x) - f(x_1)| = |x_1| > \epsilon$.

The second apparent paradox is a function that is continuous at every irrational number and discontinuous at every rational number other than 0:

$$g(x) = \begin{cases} 1/q, & \text{if } x = p/q \text{ is rational and } p \text{ and } q \text{ are relatively prime,} \\ 0, & \text{if } x \text{ is irrational or zero.} \end{cases}$$

Notice that if we select any irrational number, α, and any $\epsilon > 0$, there are only finitely many positive integers $q < 1/\epsilon$, and therefore only finitely many rational numbers p/q within one unit of α and such that $1/q > \epsilon$.

As long as we pick a $\delta > 0$ that is less than the distance to the nearest of these, then for all x within δ of α, $g(x)$ is either 0 or $1/q < \epsilon$. Therefore, $|g(\alpha) - g(x)| = |g(x)| < \epsilon$. On the other hand, every open interval contains irrational numbers, where $g(x) = 0$. If we choose any $\epsilon < 1/q$, the value of $|g(p/q) - g(x)|$ is $1/q > \epsilon$ when x is irrational.

Cauchy made it clear that he was not interested in continuity at a particular point. For him, continuity had to apply to every value in an interval. There is nothing wrong with that, but he then made an assumption that led him into a serious error.

To prove continuity on an interval, we have to show that for each value a on that interval, we can control the tolerance or allowed variation of f around the value $f(a)$, specified by $\epsilon > 0$, by controlling the permitted variation in the value of x around a, the response $\delta > 0$. The trouble is that the response δ depends on *both* ϵ and a. For example, the function defined by $h(x) = 1/x$ is continuous on the open interval $(0, 1)$ because it is continuous at each point on that interval. But if for example $\epsilon = 0.1$, then the required δ response gets smaller as x gets closer to 0. There is no single positive δ response that works for all values of x in this interval.

To prove that the definite integral exists for any continuous function, Cauchy needed something stronger than just continuity at each point in the interval. He needed what would later come to be called *uniform continuity*, meaning that for each ϵ, there is a response δ that works at *every* point of the interval. It seems that this is false, but what saved Cauchy was that he only worked with *closed* intervals. His argument worked because a function that is continuous on a closed and bounded interval is also uniformly continuous on that interval.[14]

4.5
Uniform Convergence

Uniform continuity is continuity for which the same δ response to a given $\epsilon > 0$ works for all values of x. The uniformity comes across all values of x. There is a similar concept of uniform convergence of an infinite series $\sum f_n(x)$, where the response M to a given $\epsilon > 0$ works for all values of x. Here again, Cauchy did not acknowledge this distinction. In this case, it led to a major source of confusion.

In *Cours d' Analyse*, after defining what is meant by the convergence of an infinite series of functions, the first theorem that Cauchy proves is that every infinite series of continuous functions is continuous:

THEOREM I —When the terms of a series are functions of a single variable x and are continuous with respect to this variable in the neighborhood of a particular value where the series converges, the sum $S(x)$ of the series is also in the neighborhood of this particular value, a continuous function of x. (Cauchy, 1821, pp. 131–132)

As Abel famously observed a few years later, "It appears to me that this theorem suffers exceptions."[15] As, indeed, it does. The Fourier series we encountered in equation (3.14) in section 3.6 is an infinite series of continuous functions, but it is not continuous on any interval that includes and extends beyond $(0, \pi)$.

To understand how Cauchy could prove a theorem that suffers such exceptions, it is useful to look at his proof. Given an infinite series of functions, $\sum f_n(x)$, to say that it equals $S(x)$ means that we can force the partial sums,

$$S_n(x) = f_1(x) + f_2(x) + \cdots + f_n(x),$$

as close to $S(x)$ as we want by taking n sufficiently large. In other words, given $\epsilon > 0$, there is a response M for which $n \geq M$ implies that $|S(x) - S_n(x)| < \epsilon$.

To prove continuity, we must show that we can make $S(x) - S(y)$ as close to zero as we wish by restricting the distance between x and y. This difference can be broken into three pieces:

$$S(x) - S(y) = (S(x) - S_n(x)) + \big(S_n(x) - S_n(y)\big) + \big(S_n(y) - S(y)\big).$$

Convergence of the infinite series means that we can force both $S(x) - S_n(x)$ and $S_n(y) - S(y)$ as close to zero as we wish. We make n big enough that both of these are small. Earlier, in discussing continuity, Cauchy had proven that any finite sum of continuous functions is also continuous. Since $S_n(x)$ is a finite sum of continuous functions, it must be continuous, and so we can also force $S_n(x)$ as close to $S_n(y)$ as we wish by taking x and y sufficiently close. Because all those differences can be brought arbitrarily close to zero, so can their sum.

A simple example illustrates where this can go wrong. Let

$$S_n(x) = \frac{nx^2}{1 + nx^2}.$$

If you want $S_n(x)$ to be a partial sum of infinite series, just define

$$f_1(x) = S_1(x) = \frac{x^2}{1 + x^2},$$

$$f_n(x) = S_n(x) - S_{n-1}(x)$$

$$= \frac{nx^2}{1 + nx^2} - \frac{(n-1)x^2}{1 + (n-1)x^2} = \frac{x^2}{(1 + nx^2)(1 + (n-1)x^2)}.$$

The functions $S_n(x)$ are continuous, but the limit is not continuous at $x = 0$,

$$S(x) = \lim_{n \to \infty} S_n(x) = \begin{cases} 0, & \text{if } x = 0, \\ 1, & \text{if } x \neq 0. \end{cases}$$

There is no problem keeping $S_n(0)$ close to $S(0)$, because they are both equal to 0. For $y \neq 0$, the distance from $S_n(y)$ to $S(y)$ is equal to

$$1 - S_n(y) = \frac{1}{1 + ny^2}.$$

If y is very close to 0, we will need a very large n to make this small. On the other hand, the distance from $S_n(y)$ to $S_n(0) = 0$ is

$$S_n(y) - 0 = \frac{ny^2}{1 + ny^2}.$$

With a large value of n, this is close to 1 unless we take y very, very close to 0. In fact,

$$\left(1 - S_n(y)\right) + \left(S_n(y) - 0\right) = 1.$$

There can be no choice of n and y that makes both of these small.

We *can* conclude that the infinite series is continuous if the response n does not depend on the value of y. This is precisely the condition known as uniform convergence. Today, its necessity seems obvious, but the issues were not immediately apparent to the mathematicians of the nineteenth century. Lützen has given a detailed account of the nineteenth-century struggles to clarify what is needed to complete Cauchy's proof. Both Philipp Ludwig von Seidel (1821–1896) in 1847 and George Gabriel Stokes in 1849 attempted this. Neither was entirely successful.[16]

4.6
Integration

As Cauchy came to realize that he could not assume all functions can be represented by power series, he also recognized that it was necessary to define integration in a way that did not rely on having an antiderivative. In 1823, he became the first textbook author to use a limit of sums as the starting point for the study of integration. He then proved that, with this limit definition, every function that is continuous on $[a, b]$ is also integrable on this interval. This is an incredibly powerful result: Continuity alone is enough to ensure that a function has a definite integral. The result is so important that it is worth explaining in some detail.

Cauchy needed a way of describing this *definite* integral and mentioned three alternative notations that had been used:

$$\int_{x_0}^{X} f(x)\, dx, \quad \int f(x)\, dx \left[\begin{array}{c} x_0 \\ X \end{array} \right], \quad \int f(x)\, dx \left[\begin{array}{c} x = x_0 \\ x = X \end{array} \right].$$

He immediately recognized the first, which had been suggested by Fourier, as clearly superior to the others, and so he established the notation that would be used ever after to denote a definite integral.

He then defined the definite integral as a limit of sums. For each partition of the interval $[x_0, X]$,

$$x_0 < x_1 < x_2 < \cdots < x_{n-1} < x_n = X,$$

he formed the sum

$$S = (x_1 - x_0)f(x_0) + (x_2 - x_1)f(x_1) + \cdots + (x_n - x_{n-1})f(x_{n-1}),$$

what today we would anachronistically refer to as a left Riemann sum. If this sum approaches a limit as the length of the largest subinterval shrinks to 0, then the function is *integrable* over $[x_0, X]$ and the value of the definite integral is that limit.

For a given partition, let M_i be the maximum value of f on the ith interval, from x_{i-1} to x_i, and let m_i be the minimum value of f on this interval. Cauchy's key observation was that S is sandwiched between

$$(x_1 - x_0)m_1 + (x_2 - x_1)m_2 + \cdots + (x_n - x_{n-1})m_n$$

and

$$(x_1 - x_0)M_1 + (x_2 - x_1)M_2 + \cdots + (x_n - x_{n-1})M_n.$$

If we let $V_i = M_i - m_i \geq 0$, today called the *variation* over the ith interval, then the absolute value of the difference between the upper and lower bounds on S is

$$(x_1 - x_0)(M_1 - m_1) + (x_2 - x_1)(M_2 - m_2) + \cdots + (x_n - x_{n-1})(M_2 - m_2)$$

$$(4.2) \qquad = (x_1 - x_0)V_1 + (x_2 - x_1)V_2 + \cdots + (x_n - x_{n-1})V_n.$$

Now we use uniform continuity to control the size of the variation by controlling the lengths of the subintervals in the partition. Given any $V > 0$, we can force all V_i to be strictly less than V by taking our intervals sufficiently small:

$$(x_1 - x_0)V_1 + (x_2 - x_1)V_2 + \cdots + (x_n - x_{n-1})V_n$$

$$< (x_1 - x_0)V + (x_2 - x_1)V + \cdots + (x_n - x_{n-1})V$$

$$= (X - x_0)V.$$

Because V can be any positive amount, we can narrow the range of possible values of the sum to be as small as we want. As Cauchy observed, this implies that the sums corresponding to the refinements of our original partition are approaching a single value that is defined to be the value of the definite integral.

Now Cauchy needed to tie his definition of the integral to the traditional definition: Given a function F that has f as its derivative, define $\int_a^b f(x)\, dx$ to be $F(b) - F(a)$. This connection between the limit definition of the integral and the definition using antiderivatives would, in the

1870s, come to be known as the Fundamental Theorem of Integral Calculus. Cauchy did not use this term. He did not even designate it as a theorem. He simply justified that, first, his definite integral can be used to construct a function,

$$F(x) = \int_a^x f(t)\, dt,$$

whose derivative is f and, second, that *every* function whose derivative is f differs by a constant from this constructed antiderivative F.

Today it is common to base the proof of the Fundamental Theorem of Integral Calculus on the mean value theorem for derivatives, which we discussed in section 3.5, and the mean value theorem for integrals, which asserts that if f is continuous on the closed interval $[a, b]$, then there is a value $c \in (a, b)$ such that

$$\int_a^b f(x)\, dx = (b - a) f(c).$$

The mean value theorem for integrals is an easy consequence of Cauchy's definition of the definite integral and the intermediate value theorem. If we let M be the maximum value of f on our interval $[a, b]$ and m the minimum value, then every approximating sum, S, is bounded by

$$(b - a)m \le S \le (b - a)M,$$

from which it follows that the limit of these approximating sums, the actual value of the integral, is also within these bounds:

$$(b - a)m \le \int_a^b f(x)\, dx \le (b - a)M, \quad \text{or} \quad m \le \frac{1}{b - a} \int_a^b f(x)\, dx \le M.$$

Because f is continuous on $[a, b]$, it takes on every value between m and M. In particular, there is a value c for which

$$f(c) = \frac{1}{b - a} \int_a^b f(x)\, dx.$$

Equipped with the mean value theorems, Cauchy was now ready to prove that the derivative of $\int_a^x f(t)\, dt$ is $f(x)$. Cauchy had earlier defined the derivative of F at x as the limit as α approaches 0 of

$$\frac{F(x+\alpha)-F(x)}{\alpha} = \frac{1}{\alpha}\left(\int_a^{x+\alpha} f(t)\, dt - \int_a^x f(t)\, dt\right)$$

$$= \frac{1}{\alpha}\int_x^{x+\alpha} f(t)\, dt.$$

By the mean value theorem for integrals, we can rewrite this last integral as $\alpha f(c)$ for some c between x and $x+\alpha$:

$$\frac{F(x+\alpha)-F(x)}{\alpha} = \frac{1}{\alpha}\,\alpha f(c) = f(c).$$

While c looks like a constant, it is a variable that depends on x and α and lies between x and $x+\alpha$. As α approaches 0, c approaches x. Because f is continuous, $f(c)$ converges to $f(x)$,

$$F'(x) = \lim_{\alpha\to 0}\frac{F(x+\alpha)-F(x)}{\alpha} = \lim_{\alpha\to 0} f(c) = f(x).$$

He then observed that if two functions have the same derivative, then their difference is constant. This is a consequence of the mean value theorem for derivatives: If $f'(c)=0$ for all values of c, then

$$\frac{f(b)-f(a)}{b-a} = f'(c) = 0 \quad \text{implies} \quad f(b) = f(a).$$

Because this is true for every pair a and b, f is constant. This then implies that if G is *any* function that has f as its derivative, then

$$\int_a^x f(t)\, dt - G(x) = C.$$

If we set $x = a$, the integral becomes 0, and we see that $C = -G(a)$, from which it follows that

$$\int_a^x f(t)\, dt = G(x) - G(a).$$

It was inevitable that once Cauchy had proven that every continuous function is integrable, others began asking whether discontinuous functions could be integrable. Clearly, a single point of discontinuity, say at $x = c$, is no obstacle. If (x_{i-1}, x_i) is an interval of the partition that contains c, then we cannot make the variation over that interval, V_i, smaller than the largest of $\left|\lim_{x \to c^-} f(x) - \lim_{x \to c^+} f(x)\right|$, $\left|\lim_{x \to c^-} f(x) - f(c)\right|$, and $\left|f(c) - \lim_{x \to c^+} f(x)\right|$. However, we can restrict the size of our subintervals so that the contribution of $(x_i - x_{i-1}) V_i$ is as small as we wish.

Because one point of discontinuity is not a problem, neither are two, nor any finite number of points of discontinuity. What about an infinite number of points of discontinuity? This can create problems. It was Gustav Lejeune Dirichlet who first dreamed up the function

$$f(x) = \begin{cases} 1, & \text{if } x \text{ is rational,} \\ 0, & \text{if } x \text{ is irrational,} \end{cases}$$

which is discontinuous at every point and for which the definite integral cannot exist.

But there are cases of functions with infinitely many discontinuities for which the definite integral does exist. Consider

$$g(x) = \begin{cases} 1, & \text{if } x = 1/n, \ n \in \mathbb{N}, \\ 0, & \text{otherwise,} \end{cases}$$

over the interval $[-1, 1]$. We have points of discontinuity at $x = 0$ and at $x = 1/n$ for all $n \in \mathbb{N}$, and the variation over any interval that contains at least one of these points is 1. But we can control the contribution from the subinterval that contains 0 just by limiting its length, and that subinterval will contain all but finitely many of the other points of discontinuity.

As scientists clarified the basic concepts of calculus and made their definitions precise, they were able to prove remarkable theorems. The nineteenth century would witness an expansion of the breadth and power of calculus as mathematicians deepened and strengthened its foundations. The entire subject was so transformed that continuing to call it "calculus" was no longer sufficient. By the twentieth century, mathematicians had adopted one of the terms commonly employed in the seventeenth and eighteenth centuries when referring to calculus, "analysis," to refer to this expanded understanding of the subject.

Chapter 5

ANALYSIS

In the nineteenth century, all of mathematics changed in fundamental ways. It was broadened and deepened at the same time as appreciation for the power of mathematical insights grew. And it gave rise to a profession. Universities and technical institutes sprang up, requiring faculty who could teach its advanced topics. Mathematics, which previously had been a career choice with uncertain financial prospects, now saw the creation of a solid bourgeoisie of practitioners.

Mathematics experienced an increased focus on precise definitions and careful proofs. The freewheeling style of Euler was giving way to the detailed analysis of Cauchy. Calculus became the subject we today call analysis. One thread of analysis that continued throughout this century followed questions surrounding Fourier series. In this chapter, we will explore some of the fruits of this pursuit, starting with Riemann's definition of and work on the integral and culminating with surprising insights into the true nature of the real numbers. This is only a brief sampling, intended to give a taste of what happened to calculus in this century of transformation.

5.1
The Riemann Integral

Bernhard Riemann (1826–1866) studied with Carl Friedrich Gauss (1777–1855) and Gustav Lejeune Dirichlet. Probably the most powerful mathematician of the nineteenth century, he totally revolutionized geometry and analysis and, in a single paper, laid the groundwork for the proof of the prime number theorem. This work showed how calculus over the complex numbers could be used to prove that the number of primes less than or equal to n is asymptotically[1] equal to $n/\ln n$. For his "Habilitation" of 1854, an advanced doctorate required to establish eligibility for a position as professor at a German university, Riemann chose to establish necessary and sufficient conditions for an arbitrary function to be represented by a Fourier series.

The resulting thesis, "On the Representability of a Function by a Trigonometric Series," began with a survey of the history of this problem. Riemann then established necessary and sufficient conditions for a function to be integrable. The key is that, given any prescribed upper bound $s > 0$, it must be possible to restrict the places where the variation is larger than s to a union of intervals whose total length can be made as small as we wish.

To make this more precise, we need to define the *variation of f at a point* c. Consider the variation of f over all open intervals that contain c. The variation at c, $V(c)$, is the greatest lower bound of these variations taken over the open intervals that contain c. In particular, f is continuous at c if and only if $V(c) = 0$. The function f is integrable over $[a, b]$ if and only if, for any two small values $\sigma > 0$ and $s > 0$, the points with variation greater than or equal to s can be put inside a union of open intervals whose total length is less than σ.

The proof of this theorem is made simpler by defining the definite integral as the limit of

$$\sum_{i=1}^{n}(x_i - x_{i-1})f(x_i^*),$$

where x_i^* can be any point in the interval $[x_{i-1}, x_i]$. As shown in section 4.6, the definite integral exists if and only if we can bring

(5.1) $(x_1 - x_0)V_1 + (x_2 - x_1)V_2 + \cdots + (x_n - x_{n-1})V_n$

as close to zero as we wish by controlling the length of the longest of these subintervals. Continuous functions are integrable because we can force the variation over each interval, V_i, to be as small as we wish by taking sufficiently short subintervals. But we can also force this sum to be small if

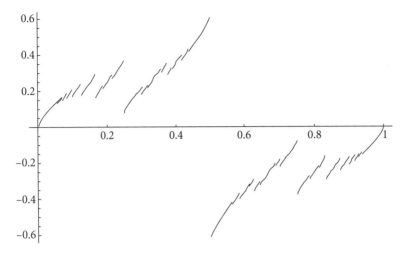

Figure 5.1. Riemann's integrable function with discontinuities in every interval, $R(x) = \sum_{n=1}^{\infty} \frac{((nx))}{n^2}$.

we can limit the sum of the lengths of the subintervals where the variation is large.

For example, a bounded function with just one point of discontinuity is integrable. Even though the variation over any interval that contains this point cannot be made any smaller than the variation at that point, we can make the interval so short that the contribution to (5.1) is as small as we wish.

Riemann's definition of the definite integral is unwieldy, but it works perfectly for the purpose to which he intended it: to establish necessary and sufficient conditions for a function to be integrable.

Riemann immediately produced a function that has a point of discontinuity inside *every* open interval, no matter how small, but which is nonetheless integrable. His function was

$$(5.2) \qquad R(x) = \sum_{n=1}^{\infty} \frac{((nx))}{n^2},$$

where $((x))$ is x minus the nearest integer unless x is exactly halfway between two integers, in which case it equals 0. For example, $((1.2)) = 1.2 - 1 = 0.2$, $((2.7)) = 2.7 - 3 = -0.3$, $((4.5)) = 0$. Although this function has a point of discontinuity inside every interval, for each value $s > 0$, there

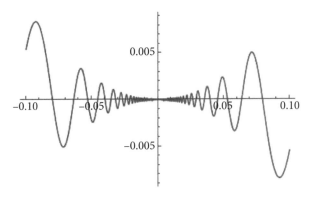

Figure 5.2. $D(x) = x^2 \sin x^{-1}$, $x \neq 0$; $D(0) = 0$.

are only finitely many points for which the variation exceeds s. A graph of this function is given in Figure 5.1.[2]

Riemann's final contribution to the definite integral was the introduction of the concept of an *improper integral*. He pointed out that it may be possible to define an integral of an unbounded function by taking a limit. As an example, $\int_0^1 x^{-1/2}\, dx$ is 2 because

$$\lim_{a \to 0^+} \int_a^1 x^{-1/2}\, dx = \lim_{a \to 0^+} 2x^{1/2}\Big|_a^1 = \lim_{a \to 0^+} 2 - 2a^{1/2} = 2.$$

While $x^{-1/2}$ is not integrable over $[0, 1]$, it does possess an improper integral.

5.2

Counterexamples to the Fundamental Theorem of Integral Calculus

As long as we stick to continuous functions, the Fundamental Theorem of Integral Calculus is valid. But if we are working with functions with infinitely many discontinuities, we can no longer assume the equivalence of integration as a limit of Riemann sums and integration as anti-differentiation. One such example comes from Riemann's function given in equation (5.2).

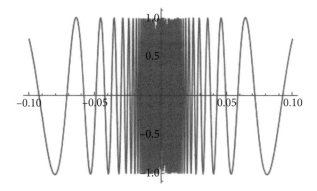

Figure 5.3. $D'(x) = 2x \sin x^{-1} - \cos x^{-1}, \quad x \neq 0; \quad D'(0) = 0.$

Functions that are derivatives are not necessarily continuous. The standard example is what I will call D, for *discontinuous derivative* (Figure 5.2),

$$(5.3) \qquad\qquad D(x) = x^2 \sin x^{-1}, \quad x \neq 0; \quad D(0) = 0.$$

When x is not zero, its derivative is $2x \sin x^{-1} - \cos x^{-1}$. At $x = 0$, one has to use the limit definition of the derivative:

$$D'(0) = \lim_{x \to 0} \frac{D(x) - D(0)}{x - 0} = \lim_{x \to 0} \frac{x^2 \sin x^{-1} - 0}{x} = \lim_{x \to 0} x \sin x^{-1} = 0.$$

But D' is not continuous at $x = 0$ because

$$\lim_{x \to 0} \left(2x \sin x^{-1} - \cos x^{-1} \right)$$

does not exist.

As Gaston Darboux (1842–1917) proved in the 1870s, any function that is a derivative must have the intermediate value property.[3] That is to say, if f is the derivative of another function, then for every pair $a < b$, and any value m between $f(a)$ and $f(b)$, there must be some $c \in [a, b]$ for which $f(c) = m$. The function D defined in equation (5.3) has a derivative that is not continuous at $x = 0$, but D' still possesses the intermediate value property: Every open interval that contains $x = 0$ also contains points where D' is equal to $+1$, -1, and every value in between (Figure 5.3).

Darboux's result implies that Riemann's integrable function in equation (5.2) cannot be a derivative. If we define

(5.4)
$$F(x) = \int_0^x R(t)\,dt,$$

then F does not have a derivative at any of the points of discontinuity of R. Every open interval in $[0, 1]$ contains infinitely many values of x at which F is not differentiable. Just because a function is defined as an integral does not imply that it can be differentiated.

What about the other direction? If a function f is known to be the derivative of another function F, is it always possible to integrate it? Strictly speaking, no unbounded function can be the limit of Riemann sums. It follows that the derivative of $x^{1/3}$ cannot be integrated over an interval that includes $x = 0$. But that is not quite fair, because the improper integral does exist. Here is a stronger version of the question: If a function F is known to be differentiable at every point in the interval $[a, b]$ and if its derivative f stays bounded over this interval, does it always follow that its derivative can be integrated over this interval? That is to say, if F' exists and is bounded on $[a, b]$, is it always true that

$$\int_a^b F'(x)\,dx = F(b) - F(a)\ ?$$

The surprising answer is "No" because the definite integral may not exist. This was established by Vito Volterra (1860–1940) when he was 20. It was published a year later, in 1881.[4] An explanation of how this function is defined and why it provides an example of a function with a bounded derivative that cannot be integrated is given in (Bressoud, 2008, pp. 89–94).

Despite these disturbing discoveries, the real problem with the Riemann integral was not the fact that differentiation and integration are not always inverse processes. The real problem was that the Riemann integral—defined to clarify the issue of when a discontinuous function could be integrated—turned out to be poorly suited to proving other results about integration. In particular, one of the important problems of the late nineteenth century was to characterize those series for which one could

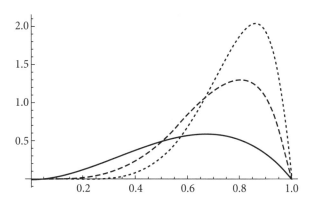

Figure 5.4. Graphs of $S_n(x) = n^2 x^n (1-x)$, $n = 2$ (solid), 4 (dashed), and 6 (dotted).

integrate the series by integrating each summand. This was of particular significance for Fourier series and the other series that arose when solving partial differential equations.

An example of a series that cannot be integrated by integrating each summand is given by

$$\sum_{k=1}^{\infty} [k^2 x^k (1-x) - (k-1)^2 x^{k-1}(1-x)]$$

whose partial sums are (see Figure 5.4)

$$S_n(x) = n^2 x^n (1-x).$$

As n increases, the hump in the graph of S_n gets pushed further to the right. For every x in $[0, 1]$, $S_n(x)$ approaches 0 as n gets larger. Therefore,

$$\sum_{k=1}^{\infty} \left[k^2 x^k (1-x) - (k-1)^2 x^{k-1}(1-x) \right] = 0, \quad 0 \le x \le 1.$$

The integral of this series is zero,

$$(5.5) \quad \int_0^1 \left(\sum_{k=1}^{\infty} \left[k^2 x^k (1-x) - (k-1)^2 x^{k-1}(1-x) \right] \right) dx = \int_0^1 0 \, dx = 0.$$

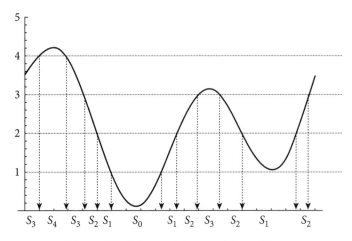

Figure 5.5. Lebesgue's horizontal partition.

But the area under $S_n(x)$ is $n^2/(n+1)(n+2)$, which approaches 1 as n gets larger:

$$(5.6) \qquad \lim_{n\to\infty} \sum_{k=1}^{n} \left(\int_0^1 \left[k^2 x^k (1-x) - (k-1)^2 x^{k-1}(1-x) \right] dx \right)$$

$$= \lim_{n\to\infty} \sum_{k=1}^{n} \left(\frac{k^2}{(k+1)(k+2)} - \frac{(k-1)^2}{k(k+1)} \right)$$

$$= \lim_{n\to\infty} \frac{n^2}{(n+1)(n+2)}$$

$$= 1.$$

In this example, the integral of the sum is not the same as the sum of the integrals.

In his doctoral thesis of 1901, Henri Lebesgue (1875–1941) proposed a different approach to integration that removed many of the difficulties associated with the Riemann integral. Rather than partitioning the domain of the function to be integrated, he chose to partition the range. In Figure 5.5, the range has been partitioned at intervals of width 1. The intervals denoted by S_1 are those where the value of the function lies between 1 and 2. We let $m(S_1)$, called the *measure* of S_1, denote the sum of the lengths of the intervals where the function lies between 1 and 2. In general, $m(S_i)$ is the sum of the lengths of the intervals where the value of the function lies between i and $i+1$. (In section 5.4 we will see the definition of measure for an arbitrary set.) We get a lower bound for the integral over $[a, b]$ by taking the sum of the measures multiplied by these lower bounds, an upper bound by taking the sum of the measures multiplied by the upper bounds

$$\sum_i i \cdot m(S_i) \leq \int_a^b f(x)\, dx \leq \sum_i (i+1) \cdot m(S_i).$$

If we take a finer partition, say at points y_i for which $y_{i+1} = y_i + \Delta y$ as i ranges over all integers, and let S_i be the set of points for which $y_i \leq f(x) < y_{i+1}$, then the definite integral is bounded below and above by

$$\sum_{i=-\infty}^{\infty} y_i \cdot m(S_i) \leq \int_a^b f(x)\, dx \leq \sum_{i=-\infty}^{\infty} y_{i+1} \cdot m(S_i).$$

The Lebesgue integral exists if and only if the difference between the upper and lower summations can be brought arbitrarily close together by taking Δy sufficiently small. The difference between the upper and lower sums is just

(5.7) $$\sum_{i=-\infty}^{\infty} (y_{i+1} - y_i) m(S_i) = \Delta y \sum_{i=-\infty}^{\infty} m(S_i) = \Delta y\, (b - a).$$

Since $b - a$ is finite, if both the upper and lower bounds on the integral are finite, then we can make this difference as small as we desire by taking Δy sufficiently close to 0. In this case, the upper and lower bounds on the Lebesgue integral approach the same value.

It is worth noting that Lebesgue's approach handles functions that are unbounded in one direction, with no need to resort to improper integrals. If the lower limit approaches $+\infty$, then the value of the definite integral is defined to be $+\infty$. If the lower limit approaches some finite value, the upper limit must also approach this value, and the definite integral has a finite value. Furthermore, if we allow S_i to be an infinite union of intervals, then

Volterra's function ceases to be a counterexample to the Fundamental Theorem of Integral Calculus. Its derivative is integrable using the Lebesgue
integral. But most significantly, the Lebesgue integral greatly simplified
the problem of deciding when a series could be integrated term-by-term.
Today, implicitly or explicitly, most mathematicians are using the Lebesgue
integral.

To make the Lebesgue integral work, we need to define what we mean
by $m(S)$ for *any* bounded subset S of real numbers, and this means that we
need to understand the possible structure of subsets of the real number line,
a challenge that turned out to be far more difficult than any mathematician
from the first half of the nineteenth century could have imagined and which
we will explore in the final section of this chapter, section 5.4.

But even the Lebesgue integral is not without problems. If we consider
the function

$$f(x) = x^2 \sin\left(x^{-2}\right), \ x \neq 0; \quad f(0) = 0,$$

it has a well-defined derivative,

$$f'(x) = 2x \sin\left(x^{-2}\right) - 2x^{-1} \cos\left(x^{-2}\right), \ x \neq 0; \quad f'(0) = 0.$$

Because this derivative is not bounded either above or below in any open
interval containing 0, its Lebesgue integral does not exist, even though
the improper Riemann integral does. Between 1912 and the 1960s, several
mathematicians created equivalent versions of a definition of the integral
that circumvents this problem, what is often referred to as the Henstock
integral.[5]

The takeaway message is that the whole issue of integration is far more
complicated than we ever let on in single variable calculus. However, students do need to be conversant in both integration as anti-differentiation
and integration as a limiting process of summation. The Fundamental
Theorem of Integral Calculus is about connecting these two views of
integration, and much of the power of calculus rests in precisely this
connection.

Figure 5.6. Karl Theodor Wilhelm Weierstrass.

5.3
Weierstrass and Elliptic Functions

One cannot discuss the development of analysis in the nineteenth century without talking about Weierstrass (1815–1897), described by Bell as "the father of analysis."[6] We met him earlier as the person who validated Euler's representation of the sine as an infinite product (section 3.3). From 1856 onward, he held a chair in mathematics at the University of Berlin, where he taught analysis on a two-year cycle and trained many of the greatest mathematicians of the late nineteenth century,[7] including Sofia Kovalevskaya (1850–1891), the first woman to be appointed to a professorship in a European university. Weierstrass was largely responsible for our modern understanding of uniform continuity and uniform convergence. He proved that if a series converges uniformly, then it can be integrated by integrating each summand,

$$\int_a^b \left(\sum_{j=1}^{\infty} f_j(x) \right) dx = \sum_{j=1}^{\infty} \left(\int_a^b f_j(x)\, dx \right).$$

Weierstrass was generous with his mathematical insights, often presenting them in class and allowing his students to refine and then publish them.

This was the case with the first example of a continuous function that cannot be differentiated at any value. Weierstrass presented this example to his class in 1872. Three years later, his student Paul du Bois-Reymond published it. An accessible account of many of Weierstrass's contributions can be found in Dunham's *The Calculus Gallery*.[8]

Weierstrass's route to success was not straightforward. His father's goal for him was to obtain an administrative position in the Prussian government. For that reason, he sent him to university to study law, finance, and economics. Frustrated that he was not allowed to pursue mathematics, Weierstrass neglected all of his courses and never bothered to take his final examinations. A year after failing college, he enrolled at the Academy in Münster to become a high school mathematics teacher.[9] In 1841, just shy of his twenty-sixth birthday, he finally graduated and took his first teaching position.

Fortunately, among Weierstrass's teachers in Münster was Christoph Gudermann (1798–1852), one of the few experts at that time in elliptic and Abelian functions. Weierstrass's greatest contributions arose from his work on this class of functions that, unfortunately, few undergraduate mathematics majors ever encounter. In his spare time, he delved into their mysteries, publishing occasional papers that received little attention until 1854 when he produced "Zur Theorie der Abelschen Functionen" (On the theory of Abelian functions), a work of such importance that the University of Königsberg awarded him an honorary doctorate and the University of Berlin hired him as a professor.

To discuss what Weierstrass accomplished requires knowledge of calculus over the complex numbers and therefore is well beyond what we could explain in these pages. Nevertheless, because elliptic functions are important, central to the most exciting mathematics of today from the proof of Fermat's Last Theorem[10] to the string theory of modern physics, it is worth giving an indication of how they are defined and why they are important.

Elliptic functions got their name from a problem that vexed Newton: to find the length of an arc of an ellipse. As mentioned in section 2.6, the formula for arc length,

$$\int_a^b \sqrt{1 + \left(\frac{dy}{dx}\right)^2}\, dx,$$

was known by 1659. Knowing that a planet's path is an ellipse naturally led to the problem of finding the arc length of a portion of an ellipse. If we consider the upper arc of an ellipse centered at the origin with $a > b > 0$,

$$\left(\frac{x}{a}\right)^2 + \left(\frac{y}{b}\right)^2 = 1, \quad \text{or}$$

$$y = b\sqrt{1 - \left(\frac{x}{a}\right)^2}$$

$$= \frac{b}{a}\sqrt{a^2 - x^2},$$

the derivative is

$$\frac{dy}{dx} = \frac{-b}{a}\frac{x}{\sqrt{a^2 - x^2}}.$$

The arc length from 0 to t is given by

$$\int_0^t \sqrt{1 + \frac{b^2}{a^2}\frac{x^2}{a^2 - x^2}}\,dx = \int_0^t \frac{\sqrt{a^2 - x^2 + b^2 x^2/a^2}}{\sqrt{a^2 - x^2}}\,dx$$

$$= \int_0^t \frac{a^2 - \left(1 - \frac{b^2}{a^2}\right)x^2}{\sqrt{\left(a^2 - \left(1 - \frac{b^2}{a^2}\right)x^2\right)(a^2 - x^2)}}\,dx$$

$$= \int_0^t \frac{a^2 - k^2 x^2}{\sqrt{(a^2 - k^2 x^2)(a^2 - x^2)}}\,dx,$$

where $k^2 = 1 - b^2/a^2$.

The problem comes from that square root of a bi-quadratic polynomial[11] in the denominator of the function to be integrated. Other similar integrals were discovered at the same time, most famously the integral needed to determine when a simple pendulum would reach a given point along its arc.[12] These integrals—with an integrand that consists of a polynomial in the numerator and the square root of a cubic or bi-quadratic polynomial in the denominator—came to be known as *elliptic integrals*. The case where the denominator is the square root of a polynomial of degree greater than four are called *Abelian integrals*, named for Abel's work on them.

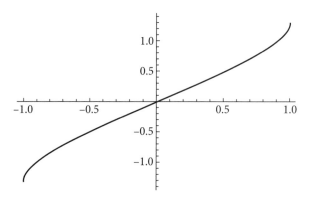

Figure 5.7. Graph of $F(t) = \int_0^t \frac{1}{\sqrt{1-x^4}}\,dx$ (over $-1 < t < 1$).

In 1797, Gauss published the first real insight into these integrals,[13] focusing on the simplest case (see Figure 5.7 for a graph of this function),

$$F(t) = \int_0^t \frac{1}{\sqrt{1 - x^4}}\,dx.$$

Gauss noted that similar functions that we can integrate—those with a square root of a quadratic polynomial in the denominator—are the inverses of more common functions,[14]

$$\int_0^t \frac{1}{\sqrt{1 - x^2}}\,dx = \arcsin t,$$

$$\int_0^t \frac{1}{\sqrt{1 + x^2}}\,dx = \operatorname{arcsinh} t,$$

where $\sinh t$ is the hyperbolic sine, $\sinh t = (e^t - e^{-t})/2$, and $\operatorname{arcsinh} t$ is its inverse function. The first insight was that rather than focusing on the function defined by the elliptic integral, we need to focus on the inverse. The *elliptic function* $E(t)$ is defined as the inverse of $F(t)$.

The second insight came from the realization that this function only reveals its true nature when it is defined on the complex plane, $\mathbb{C}$. While the sine and the hyperbolic sine appear to be very different when viewed as functions over the real number line, the distinction disappears when we look at them over the complex plane. Thanks to Euler's formula,

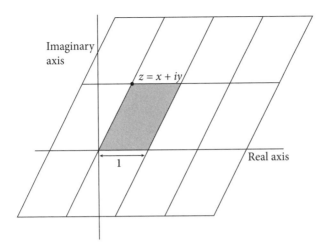

Figure 5.8. The lattice for an elliptic function with periods 1 and $z = x + iy$.

equation (3.9) in section 3.3,

$$\sin x = \frac{e^{ix} - e^{-ix}}{2i} = -i \sinh(ix).$$

As a mapping from the complex plane to itself, the hyperbolic sine is simply the ordinary sine with both domain and range rotated by 90°. In particular, over the complex plane they are both periodic functions. The sine has a real period, 2π. The hyperbolic sine has a purely imaginary period, $2\pi i$.

Elliptic functions have *two* periods. Two independent vectors in the complex plane define a parallelogram, which can be used to produce a lattice (Figure 5.8). Just as the sine is uniquely determined over the entire real line by its values over $[0, 2\pi]$, so an elliptic function is uniquely determined over the entire complex plane by its values in this parallelogram. In fact, the sine and hyperbolic sine are simply extreme cases of elliptic functions in which one of the two periods has been stretched out to infinity.

The beauty and the power of elliptic functions arise from the intricate identities and relationships that they possess. The identities for trigonometric functions are a pale shadow of what happens in the full complexity of elliptic functions. No one had a more intuitive feel for these than the Indian mathematician Srinivasa Ramanujan (1887–1920), who failed college not once but twice. As a clerk in Madras,[15] he had access to the mathematics library at the University of Madras where he learned about elliptic functions and began his own explorations. His discoveries were recognized in his brief lifetime when he became one of the youngest Fellows of the Royal Society and only the second Indian to earn this honor. Because the symmetries arising from elliptic functions are found throughout the natural world, Ramanujan's results have become foundational to much of modern physics.

5.4
Subsets of the Real Numbers

Georg Cantor is best known for his work on sets and the structure of the real numbers, but he began with a question about Fourier series. Cantor had studied number theory with Kummer and Weierstrass at the University of Berlin. His first job after completing his "Habilitation" was at the University of Halle-Wittenberg where Eduard Heine convinced him to work on the problems that still existed with Fourier series. Cantor was soon wrestling with the Fourier series expansions of functions with infinitely many points of discontinuity. This led him to realize that not all infinite sets of real numbers are comparable.

In fact, as the mathematical community slowly came to realize, there are three very distinct ways of describing the size of a bounded, infinite subset of the real numbers: density, cardinality, and measure.

Density is the oldest and was well understood by the mid-nineteenth century. A subset D of $[0, 1]$ is *dense* in $[0, 1]$ if every open interval that overlaps $[0, 1]$ contains at least one point of D. In fact, once you know that every open interval contains at least one point of D, it is not hard to see that every open interval contains *infinitely* many points of D.[16] The classic example of a dense subset of $[0, 1]$ is the set of rational numbers in this interval. Many smaller sets, such as the set of rational numbers whose denominators are a power of 2, are also dense.

At the opposite extreme are sets that are *nowhere dense*. A set N is nowhere dense in $[0, 1]$ if every open interval contains a subinterval without any points in N. Any finite subset is nowhere dense, and so is $\{1/n \mid n = 1, 2, 3, \ldots\}$. Any open subinterval $I \subset [0, 1]$ contains a point a that is not the reciprocal of an integer, and so it lies strictly between $1/(n + 1)$ and $1/n$ for some positive integer n. The intersection of I and $\left(\frac{1}{n+1}, \frac{1}{n}\right)$ is a subinterval of I that does not contain any points of the form $1/n$.

It was Cantor who discovered the importance of the *cardinality* of infinite sets in 1873 (published the next year).[17] Two sets have the same cardinality if and only if they can be put in one-to-one correspondence. In this sense, the set of rational numbers in the interval $[0, 1]$ is no larger than the set of positive integers because we can put them in one-to-one correspondence. Starting with 0/1 and 1/1, we can put the rational numbers in a linear order: In reduced form, a/b comes before c/d if $a+b < c+d$ or if $a+b = c+d$ and $a < c$. The list of rational numbers in $[0, 1]$ with their correspondence to the positive integers is given by

$$
\begin{array}{ccccccccccccc}
 & 1 & 2 & 3 & 4 & 5 & 6 & 7 & 8 & 9 & 10 & 11 & \cdots \\[4pt]
 & \updownarrow & \updownarrow & \updownarrow & \updownarrow & \updownarrow & \updownarrow & \updownarrow & \updownarrow & \updownarrow & \updownarrow & \updownarrow & \\[4pt]
 & \frac{0}{1} & \frac{1}{1} & \frac{1}{2} & \frac{1}{3} & \frac{1}{4} & \frac{2}{3} & \frac{1}{5} & \frac{1}{6} & \frac{2}{5} & \frac{3}{4} & \frac{1}{7} & \cdots \\[4pt]
a+b: & 1 & 2 & 3 & 4 & 5 & 5 & 6 & 7 & 7 & 7 & 8 & \cdots .
\end{array}
$$

Sets that are finite or can be put in one-to-one correspondence with the set of positive integers are called *countable*. The rational numbers are countable. This may not be surprising. After all, is there not just one infinity? Cantor's 1874 paper showed that there are larger infinities. In particular, the real numbers in $[0, 1]$ *cannot* be put into one-to-one correspondence with the positive integers. The standard proof of this fact, which relies on the representation of each real number as an infinite decimal, is well-known.[18] Dunham has given a nice account of Cantor's original proof, which built directly on the completeness of the real numbers.[19] If a set is not countable, it is called *uncountable*. The set of real numbers in $[0, 1]$ is uncountable.

We encountered the third way of describing the size of a set, called *measure*, in section 5.2. Lebesgue defined it to satisfy three criteria:

(1) The measure of an interval is the length of that interval, the measure of a single point is zero, and the measure of any finite or countable union of pairwise disjoint sets with well-defined measures is the sum of the measures of those sets.

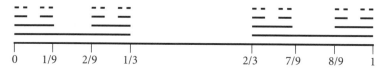

Figure 5.9. Construction of Cantor's set by removal of middle thirds.

(2) Translating a set (adding the same amount to each element) does not change its measure.

(3) If sets S and T both have well-defined measures, then so does $S \cap T$, and $S - T$ (the set of elements of S that are not in T) has well-defined measure equal to the measure of S minus the measure of $S \cap T$.

As Lebesgue was able to demonstrate, these conditions uniquely determine a way of measuring the size of subsets of the real numbers. To find the measure of a set S, we define a *cover*, C, of S to be any countable union of open intervals whose union contains S and the *length* of this cover to be the sum of the lengths of these intervals. *If the measure of S exists,*[20] then it is equal to the greatest lower bound of the set of lengths of all covers of S. If we consider subsets of $[0, 1]$, then the set of rational numbers on this interval has measure 0 (it is a countable union of sets of measure 0), and the set of irrational numbers has measure 1. Any countable set necessarily has measure 0. What about uncountable subsets of $[0, 1]$?

As Cantor showed, an uncountable subset could have measure 0. If we start with the interval $[0, 1]$ and remove the open interval $(1/3, 2/3)$, we are left with a set of measure 2/3. If we now remove the middle third of each of the remaining intervals, $(1/9, 2/9)$ and $(7/9, 8/9)$, we are down to a set of measure 4/9. We continue, at each step removing the middle third from each of the remaining intervals. After the kth step, we are left with 2^k intervals whose total measure is $(2/3)^k$ (see Figure 5.9). The set C, sometimes referred to as "Cantor dust," consists of all the points in $[0, 1]$ that are never removed. What has been removed is a countable union of open intervals, so the Cantor dust is measurable, and its measure is

$$m(C) = 1 - \sum_{k=1}^{\infty} \frac{2^{k-1}}{3^k} = 1 - \frac{1}{3}\left(\frac{1}{1 - 2/3}\right) = 0.$$

The set C clearly contains all the endpoints of our intervals, those rational numbers with a denominator that is power of 3. Perhaps surprisingly, there are uncountably many other points that are never removed.

The easiest way to see this is to consider the base 3 or *ternary* representation of the real numbers between 0 and 1. For example,

$$0.211021_3 = \frac{2}{3} + \frac{1}{3^2} + \frac{1}{3^3} + \frac{0}{3^4} + \frac{2}{3^5} + \frac{1}{3^6} = \frac{601}{729}.$$

The endpoints of the intervals are those real numbers with a finite number of digits in the base 3 representation. Each real number between 0 and 1 has a possibly infinite decimal representation using the digits 0 through 9, and the representation is unique except that a finite decimal also can be represented with an infinite string of 9s:

$$\frac{8}{10} = \frac{7}{10} + \frac{9}{10^2} + \frac{9}{10^3} + \frac{9}{10^4} + \cdots, \quad 0.8 = 0.7999\ldots.$$

In similar manner, every real number between 0 and 1 has a possibly infinite ternary representation using the digits 0, 1, 2 that is unique except that a finite representation also can be represented with an infinite string of 2s

$$\frac{1}{3} = \frac{0}{3} + \frac{2}{3^2} + \frac{2}{3^3} + \frac{2}{3^4} + \cdots, \quad 0.1_3 = 0.0222\ldots_3.$$

The Cantor dust consists of those points left over after removing the intervals $(0.1_3, 0.2_3)$, then $(0.01_3, 0.02_3)$ and $(0.21_3, 0.22_3)$, then $(0.001_3, 0.002_3)$, $(0.021_3, 0.022_3)$, $(0.201_3, 0.202_3)$, and $(0.221_3, 0.222_3)$, and so on. In other words, we remove all real numbers whose only ternary representation has a nonzero digit anywhere after a 1. A real number whose ternary representation consists only of 0s and 2s will be left in the Cantor dust. In particular, one element of the Cantor dust is

$$0.2020202\ldots_3 = \frac{2}{3} + \frac{2}{3^3} + \frac{2}{3^5} + \frac{2}{3^7} + \cdots = \frac{2}{3}\left(\frac{1}{1 - 1/9}\right) = \frac{3}{4}.$$

How many of them are there? There is an obvious one-to-one correspondence between elements of C and numbers in binary representation such as

$$0.1010101\ldots_2 = \frac{1}{2} + \frac{1}{2^3} + \frac{1}{2^5} + \frac{1}{2^7} + \cdots = \frac{1}{2}\left(\frac{1}{1-1/4}\right) = \frac{2}{3}.$$

But *every* real number between 0 and 1 has such a binary representation, so the cardinality of C is the same as the cardinality of the entire interval $[0, 1]$.

This set C is counterintuitive. It is nowhere dense: Every open interval overlapping $[0, 1]$ must intersect at least one of the open intervals that have been removed. It is uncountable. And it has measure 0.

Can a nowhere dense set have positive measure? Yes. Instead of removing middle thirds, remove a middle fifth. Every open interval will still intersect at least one of these, but the measure of what is left is

$$1 - \sum_{k=1}^{\infty} \frac{2^{k-1}}{5^k} = 1 - \frac{1}{5}\left(\frac{1}{1-2/5}\right) = \frac{2}{3}.$$

By taking smaller fractions, we can bring the measure of the remaining nowhere dense set as close to 1 as we wish.

What about a dense set, can it have measure 0? Clearly yes if it is countable, for example, the rationals. But also yes if it is uncountable. Start with the Cantor set C and put a copy of C inside the interval from 1/3 to 2/3. Then put another copy of C inside each of the three missing intervals of length 1/9. Then another copy inside each of the nine missing intervals of length 1/27. Keep doing this. The union of all of these sets is a countable union of sets of measure 0, so it also has measure 0, but it is dense in $[0, 1]$.

Of the three ways of thinking of the size of an infinite subset of $[0, 1]$:

nowhere dense $\longleftrightarrow$ dense,

countable $\longleftrightarrow$ uncountable,

measure zero $\longleftrightarrow$ positive measure,

and the eight possible ways of combining them, only two cannot occur, the two that pair "countable" with "positive measure."

5.5
Twentieth-Century Postscript

As the twentieth century progressed, it became increasingly clear that the real numbers can be strange and mysterious. Two questions arose from the measures of size described in the previous section, both of which would turn out to have surprising answers.

The first question came from Cantor's further explorations of the cardinality of infinite sets. The concept of cardinality, based as it is on one-to-one correspondences, is full of surprises. We can use the linear function $y = \pi(x - 1/2)$ to establish a one-to-one correspondence between $(0, 1)$ and $(-\pi/2, \pi/2)$. The tangent function, $\tan x$, then produces a one-to-one correspondence between $(-\pi/2, \pi/2)$ and the entire real number line. It follows that the cardinality of the entire real line is the same as the cardinality of the interval $(0, 1)$. Adding one or two points does not change an infinite cardinality, so $(0, 1]$ and $[0, 1]$ also share this cardinality, which is usually written as c for *continuum*. Because the real number line is a countable union of sets of the form $(n, n + 1]$, *any* countable union of sets of cardinality c must also have cardinality c.

The cardinality of the entire plane is also equal to the cardinality of the real numbers. Given a point with x and y coordinates between 0 and 1, say (a, b) where $a = 0.a_1a_2a_3 \ldots$ and $b = 0.b_1b_2b_3 \ldots$, we can construct a unique[21] decimal in $(0, 1)$ of the form $0.a_1b_1a_2b_2a_3b_3 \ldots$. Because the plane is made of countably many copies of the open unit square, plus countably many lines, it must have the same cardinality as the real number line.

But there are larger cardinalities. Consider the set of subsets of the real numbers. We can think of this as all possible color schemes where either blue or red is used to color each real number. If a given real number is blue, we include it in our subset. If red, we do not. If this set of possible color schemes had cardinality c, then we could set up a one-to-one correspondence between the real numbers and these color schemes. We assume we have accomplished this. For each real number α, we have a unique color scheme S_α, specifying the color of each real number. Because α is a real number, it is assigned a color by S_α. We note its color in this scheme. Now consider a color scheme, T, so that for each real number α, the color of α in T is the opposite of its color in S_α. For example, if S_π is the color scheme that corresponds to the real number π, and if π is blue in S_π, then

π is red in T. If we really have a one-to-one correspondence between color schemes for the real numbers and color schemes for the real numbers, then the color scheme T corresponds to some real number, say $T = S_\beta$. Now we have a problem because the color of β in T must be the opposite of its color in S_β, but these color schemes are the same. The assumption that there is a one-to-one correspondence must be false.

This larger cardinality we have just described is usually written 2^c. It is the cardinality of the set of mappings from the real numbers into a two-element set.[22] We can continue to find larger cardinalities by iterating this process. Given any set S, the cardinality of the set of mappings from S to a two-element set is always strictly larger than the cardinality of S. In fact, there are uncountably many larger cardinalities. What about other cardinalities that are smaller than c? Specifically, mathematicians asked if there were any infinite subsets of $[0, 1]$ with a cardinality other than countable or c.

Cantor believed that there were none, and this belief came to be known as the *continuum hypothesis*. Certainly, if such a set were to exist, no one had the slightest idea how to construct it. When, in 1900, David Hilbert (1862–1943) inspired the mathematical community with his twenty-three challenge problems for the twentieth century, the first challenge he presented was to settle the continuum hypothesis.

In 1940, Kurt Gödel (1906–1978) showed that the standard set of axioms that define the real numbers, known as the Zermelo-Fraenkel axioms, are consistent with the continuum hypothesis. This does not prove that the continuum hypothesis is true, just that one cannot disprove it or construct a counterexample within the framework of these axioms. So far, so good. But then in 1963 Paul Cohen (1934–2007) proved the the Zermelo-Fraenkel axioms are consistent with the existence of an infinite subset of $[0, 1]$ with a cardinality other than countable or c. The continuum hypothesis could be false. This does not simply mean that we know nothing about the status of the continuum hypothesis. It means that we *can* know nothing about its status without making assumptions that go beyond Zermelo-Fraenkel. More precisely, the existence of a subset of $[0, 1]$ with a cardinality other than countable or c is consistent with everything else we know about the real numbers, and the lack of such a subset is consistent with everything else we know about these numbers. You get to choose whether you want the continuum hypothesis to be true or false.[23]

Another question that surfaced in the early twentieth century concerned the existence of the Lebesgue integral of a bounded function over a bounded interval. Equation (5.7) in section 5.2 appears to confirm that we can bring the upper and lower approximating sums as close together as we wish by taking Δy sufficiently small, and therefore there must be a well-defined value for the integral. The only potential problem arises from the need to use $m(S_i)$, the measure of the set S_i. Do all subsets of $[0, 1]$ have a well-defined measure?

In 1905, Guiseppe Vitali (1875–1932) showed how to construct a non-measurable subset of $[0, 1]$.[24] There was just one difficulty with this construction. He needed to choose one representative from each of an uncountable number of sets. Making an uncountable number of choices is problematic because there is no natural way to specify a choice for each real number. A vigorous debate unfolded in the early twentieth century over whether or not such a choice was to be allowed. Allowing uncountably many choices came to be known as the *Axiom of Choice*. The debate was vigorous because this axiom implied many useful results.[25]

On the other hand, this axiom implied some very strange results. In 1924, Banach and Tarski demonstrated how the axiom of choice could be used to decompose a solid ball into five sets and then reassemble those five sets using only rigid motions (rotations and translations) into two solid balls, each of exactly the same size as the original. Volume is not preserved because volume is just a special case of measure, and not all of the five sets are measurable. More than this, their argument can be extended to show that *any* solid object can be cut into finitely many pieces and reassembled using rigid motions into any other solid object. A delightfully accessible proof of this result is contained in Wapner's *The Pea and the Sun* (2005), the point being that if we accept the axiom of choice, then it is possible, in theory, to take a pea, decompose it into finitely many pieces, and use rigid motions to reconstitute those pieces into a sphere the size of the sun.[26]

Like the continuum hypothesis, we can accept or reject the axiom of choice without otherwise affecting anything else we know about the real numbers, including whether we have chosen to accept the continuum hypothesis. The real numbers have revealed themselves to be truly bizarre.

Appendix

REFLECTIONS ON THE
TEACHING OF CALCULUS

Teaching Integration as Accumulation

It is deeply unfortunate that for many students integration is limited to finding areas and volumes and the memorization of rules for antiderivatives. There are many reasons that the failure to appreciate integration as accumulation is an impoverished view. Three stand out.

First, as history illustrates, accumulation is an intuitive process. We see this from how broadly it has been known. The ancient Egyptians almost certainly were using some form of accumulating incremental changes when they discovered the formula for the volume of a truncated pyramid. The Chinese had mastered Cavalieri's approach to finding volumes by the fifth century of the Common Era.[1] This is an aspect of calculus that clearly transcends culture.

Second, students will always have access to technologies that can find definite and indefinite integrals. While many of the techniques of integration are important for the structural insights they provide, finding antiderivatives is a skill that few students will need outside of a calculus class. Far more useful is the ability to transform an accumulation problem into a definite integral.

Third, failure to appreciate the integral as an accumulator robs students of a rich vein of applications that reveal integration as more than finding areas and volumes. The integral is a tool that can be used to study anything that accumulates at variable rates: distance traveled, work accomplished, profits earned, materials produced, and the tracking of environmental degradation or recovery.

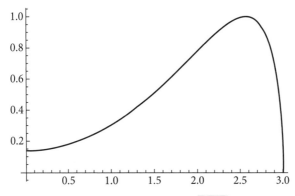

Figure A.1. The graph of $y = \sin \sqrt{9 - t^2}$, $0 \le t \le 3$.

A strong case can be made for starting calculus with an introduction to integration via accumulation. This is the approach that Thompson at Arizona State has adopted.[2] He has built his curriculum on an essential insight, that calculus is the study of functional relationships between variables that *vary*. An accumulator must be a function, describing how much has accumulated at each value of the dependent variable. The definite integral must make its first appearance as a function with a variable upper limit.

Oehrtman at Oklahoma State[3] has developed a series of exercises that provide an entry to the integral as accumulator while giving students the opportunity to construct their understanding of the principles that are involved.

He presents students with a NASA Lunar Rover that can travel up to three hours on a single charge and whose velocity at time t is given by $\sin \sqrt{9 - t^2}$ miles per hour (Figure A.1). As students explore how far it can travel away from base and still be able to return, and at what time it must turn around, they learn to estimate the distance traveled over small intervals of time. As is apparent from the graph, the velocity is increasing for a bit more than the first $2\frac{1}{2}$ hours. More precisely, it hits its maximum value when $\sqrt{9 - t^2} = \pi/2$, or approximately $t = 2.56$ hours. Students quickly catch on that when the velocity is increasing, they can underestimate distance traveled by taking left-hand values, overestimate with right-hand values, and so obtain upper and lower bounds between which the true distance must lie. A simple spreadsheet can suffice to get reasonable approximations. Students realize that they can narrow the distance between these limits by taking shorter intervals of time. Along the way, they are introduced to summation notation as convenient shorthand for what they are already doing. They are shown the definite integral with elapsed time as a variable upper limit as a way of encoding the actual accumulation function. Because their first encounter with the definite integral occurs as a representation of the accumulation function, such a function has meaning for them.

By insisting that students record upper and lower bounds with each choice of time interval, Oehrtman is also seeding the ground for the eventual introduction of limits. As we saw in chapter 4, limits in their modern sense are defined in terms of inequalities. To assign a value to

$$\int_0^2 \sin\left(\sqrt{9 - t^2}\right)\, dt$$

is to assert that given any two numbers, one greater and one less than this value, we can bring the time intervals short enough that the assigned value lies between these bounds. That, and only that, is what is meant when we say that a definite integral is the limit of Riemann sums.

Throughout this book, I have insisted on referring to the Fundamental Theorem of Integral Calculus, rather than simply the Fundamental Theorem of Calculus, for the theorem that most students remember—if they remember it at all—as connecting integration and differentiation. As I explain in the footnote in section 2.7, there is a historical justification for this. More importantly, there is a strong pedagogical reason. Most students quickly forget the limit definition of the integral. Given the emphasis that most courses place on techniques of integration and how little they use the limit definition, it is not surprising that most students view integration and reversing differentiation as synonymous. The result is that the theorem which should be a centerpiece of calculus is reduced to a tautology.

As we saw in section 4.6, Cauchy was the first to prove this theorem, and he did so in order to connect the two definitions of integration, as a limit of sums and as an antiderivative. Calling it the Fundamental Theorem of Integral Calculus is both more accurate and serves as a reminder that this theorem is really about connecting these two understandings of integration. This can help remind students that integration is not simply anti-differentiation.

Teaching Differentiation as Ratios of Change

One can question whether Indian astronomers of the first millennium had understood the derivative of the sine in anything like a modern sense. Nevertheless, they were studying ratios of change, seeking to understand how small changes in the input were reflected by changes in the output. They had discovered that, in the case of the sine, the limiting value of this ratio is an easily computed quantity that can be used to estimate the change in the output. This is a far more intuitive introduction to derivatives than slopes of tangent lines. In addition, it prepares the ground for the eventual application of the derivative to situations in which input values can only be known approximately and we need to bound the possible values of the output.

We know that even calculus students struggle with ratios, but once they understand ratios of change, they are positioned for a more solid understanding of the derivative as the slope of the tangent line. When I teach the first semester of calculus at Macalester, most of my students have already seen some calculus. For over two decades, I have begun the semester with a simple assessment of what they know. The first question asks for the instantaneous rate of change of $x^3 - 7$ at $x = 2$ and the average rate of change of $x^3 - 7$ over the interval from $x = 2$ to $x = 3$. Everyone who has seen any calculus can do the former. Almost none of these students correctly answer the latter. Many average the instantaneous rates of change at $x = 2$ and $x = 3$. Given historical difficulties with the concept of instantaneous velocity, it is ironic but true that students are more comfortable with instantaneous velocities than they are with average velocities.

I believe that several things are happening. One is that average rate of change is not a topic usually emphasized in precalculus. While it appears in the introductory material of every calculus class, it is quickly forgotten as attention turns to the many techniques of differentiation and easy ways of determining instantaneous rate of change. Another is that this is called an "average." It does not look like any average students would have studied in the elementary or middle grades. And finally, the average rate of change is a ratio.

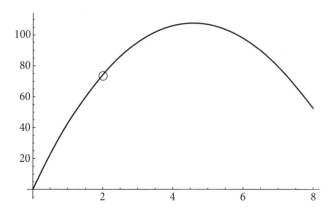

Figure A.2. The graph of $h(t) = 7350 - 245t - 7350e^{-t/25}$.

Students have difficulty grasping the significance of the limit definition of the derivative. If we try to explain to students that

$$\frac{f(x+h) - f(x)}{(x+h) - x}$$

is the slope of a secant line with one end at $(x, f(x))$ and that this becomes the slope of the tangent line in the limit as h approaches zero, the significance of what we have said is frequently lost in the midst of all these unfamiliar concepts that tumble after each other.

Instead, we can begin with a function that describes accumulation at time t, and ask about the rate at which this accumulation takes place. Oehrtman has students find an approximation to the speed of a crossbow bolt two seconds after it is fired, given that it is fired straight up into the air (do not attempt this yourself) with a height at time t given by (Figure A.2)[4]

$$h(t) = 7350 - 245t - 7350e^{-t/25} \text{ meters per second.}$$

They are asked to find underestimates and overestimates for this velocity that differ by at most 0.1 m/s, thus enabling them to both approximate the velocity and give an error bound for their approximation.

The definition of the derivative is important for understanding how to approximate a ratio of changes or a ratio of rates of change. To the extent that any theorems about the derivative are proven in the first year of calculus, the definition is essential. But students also have an intuitive appreciation for the derivative as an instantaneous rate of change, how fast an object is moving at a given instant of time. This provides a natural introduction to another side to differentiation emphasized in chapter 2, differential equations.

One of the reasons for including Napier's development of logarithms was to highlight his work on relating rates of change. In effect, he established that if y is a logarithm of x, then

$$\frac{dy}{dt} = \frac{c}{x}\frac{dx}{dt},$$

where the constant c depends on the base of the logarithm. Unfortunately, too few calculus classes convey the power and importance of differential equations. I love the story of Maxwell's equations because it illustrates one reason why we care so much about calculus, the power of mathematical models to reveal unexpected insights into our world. Many of the more innovative calculus curricula, including some of the earliest calculus reform efforts and the course we now teach at Macalester,[5] begin with differential equations and emphasize throughout that calculus is about modeling dynamical systems. Again, technology makes it easy to explore models of population growth, the spread of epidemics, and predator-prey scenarios. This provides excitement around the study of calculus. Analysis of how numerical approximations are accomplished provides a bridge back to the derivative as the limit of ratios of change.

The derivative is a rich concept that opens many doors. How unfortunate that for many students the only lasting takeaway is that it turns x^3 into $3x^2$.

Teaching Series as Sequences of Partial Sums

Too often, differentiation is reduced to finding derivatives, integration to finding antiderivatives, and series to determining convergence, a dismal view of these summations. Because few students will remember the convergence tests that they have memorized, I fully endorse the position now taken by many colleges and universities of postponing a discussion of questions of convergence of power series until students see analysis, focusing instead on the partial sums, the Taylor polynomials.[6] Though challenging, I would love to include a discussion of the Lagrange remainder theorem as a means of bounding the error. This would also convey the practical importance of the mean value theorem.

I included Euler's derivation of the power series expansion of the exponential function as an example of the kind of playfulness that infinite summations invite and that I wish our students could appreciate. It is not unreasonable to begin a student exploration with the formula for compounding interest,

$$A = P\left(1 + \frac{r}{n}\right)^{nt},$$

ask students to expand it using the binomial theorem, and then explore what happens to this expansion as n increases, letting them discover the connection between this formula and the exponential function.

While general power series are less important than Taylor polynomials at this stage in a student's career, both series of constants and the geometric series are extremely important. Geometric series are ubiquitous throughout mathematics, and when students do need to learn rules of convergence, they are the bedrock on which many convergence tests are built.

Series of constants are important because our modern understanding of limit arose from the eighteenth-century struggle to understand what it means for them to converge. These scientists realized that the critical question was whether they could bound the distance between the partial sums and the asserted value. In line with the historical development, many textbooks begin the study of limits with examples taken from infinite summations.

In addition, students need to be aware of how audacious it was when Leibniz stated that

$$\frac{\pi}{4} = 1 - \frac{1}{3} + \frac{1}{5} - \frac{1}{7} + \cdots.$$

Here is an opportunity for students to engage with the question of what such an equality means. A discussion of what is meant by $0.999\ldots$ fits naturally within this framework.

Teaching Limits as the Algebra of Inequalities

While giving first-year calculus students the ϵ-δ definition of a limit and expecting them to absorb it is irresponsible, the idea that sits behind this formalization is accessible. Whether working with integrals, derivatives, or series, they are defined by means of approximations. The limit is that prescribed value for which given *any* two bounds, one greater and one less than this prescribed value, all of the approximations lie between these bounds once we have restricted the approximations—restricted by interval length for integrals or derivatives or by minimal number of terms in the case of partial sums of a series.

The idea of limits as used in calculus may seem simple, but it is in fact remarkably complex. Explorations of student understanding have revealed that one of the most common metaphors used by students as they grapple with the idea of limit is what researchers refer to as the "collapse" metaphor. Whether explicitly or implicitly, students interpret

$$\lim_{x \to a} f(x) = L$$

to mean that as x gets closer and closer to a, $f(x)$ is getting closer and closer to L until, at the moment when x gets to a, $f(x)$ collapses onto L.

For most applications of the limit, this is not too bad and often proves valuable, but it does promote dangerous misconceptions. As Swinyard[7] has demonstrated, this is an "x-first" perspective on the limit, looking at how the action of the independent variable affects the dependent variable. The problem is that the true definition of limit is "y-first," choosing a permitted tolerance around the y-value and then establishing that there are bounds on the x-value that will ensure this. Swinyard and Larsen[8] have shown that students exhibit great difficulty comprehending the correct definition until they have shifted to a y-first understanding.

The collapse metaphor also sits behind student difficulty with a problem that British mathematician David Tall posed to 22 senior mathematics majors:[9]

If $\lim_{x \to a} f(x) = b$ and $\lim_{y \to b} g(y) = c$,

does it necessarily follow that $\lim_{x \to a} g(f(x)) = c$?

Even after repeated requests to reconsider their answers, 21 of these 22 students insisted that $\lim_{x \to a} g(f(x))$ must equal c.

It is the focus on "getting closer" that led them astray. Understanding that $f(x)$ is getting closer to b naturally leads to the assumption that $f(x)$ can stand in for y, the variable that is getting closer to b. But what if $f(x) = b$, a constant function, whereas g is discontinuous at b with the limit existing, but $\lim_{y \to b} g(y) \neq g(b)$? In this case,

$$\lim_{x \to a} g(f(x)) = g(b) \neq c.$$

A detail that students often overlook in the interpretation of $\lim_{x \to a} f(x)$ is that x cannot equal a but $f(x)$ can equal the limiting value. Notice that this sits at the heart of Tall's example. When we write $\lim_{y \to b} g(y) = c$, we are considering a variable, y, that is explicitly excluded from equaling b. But writing $\lim_{x \to a} f(x) = b$ in no way excludes the possibility that $f(x) = b$ at any or all values of x near a.

It is natural to ask why we exclude the point a. This comes from the need to use the language of limits in cases where the function in question is not defined at a, in particular in defining the derivative. The average rate of change,

$$\frac{f(x) - f(a)}{x - a},$$

is not well-defined when $x = a$. We need a definition of limit that excludes the possibility that x could equal a.

Oehrtman's examples of explorations that lead students to discover the principles of differentiation and integration grew out of his analysis of the metaphors that students use to explain limits.[10] He discovered that the language of approximations was both natural for many students and highly productive. With this in mind, he has developed tasks that force students to systematize their ideas of approximation, using the algebra of inequalities to prescribe conditions that bound the error when approximating a derivative or integral. This equips students with a meaningful understanding of these concepts and lays the foundation for an eventual transition to formal definitions.

As Oehrtman documented in 2008,[11] there are five key questions that students must answer in each context, whether differentiation, integration, or series:

(1) What are you approximating?
(2) What are the approximations?
(3) What are the errors?
(4) What are the bounds on the errors?
(5) How can the error be made smaller than any predetermined bound?

As Oehrtman explains, the last two are intentionally reciprocal: Given a description of the parameters to be used in the approximation, what are the bounds on the error? Given bounds on the error, what parameters should be used in the approximation?

Oehrtman has not been alone in taking this approach. Lax and Terrell, who begin *Calculus with Applications* with a careful introduction to limits, start with a review of inequalities.[12]

Finally, I should say something about infinitesimals. Over the millennia, they have been fruitful sources of insight. They still have great intuitive appeal, often helping working scientists to translate accumulation problems into definite integrals and to create differential equations. Although calculus based on infinitesimals can be rigorously justified, that is a twentieth-century result requiring sophisticated set theory. But the main problem with relying entirely on infinitesimals is that the transition to a modern understanding of limits becomes more difficult than it needs to be. Given how easily students grasp the use of approximations and inequalities, this seems a more natural and productive approach.

The Last Word

I hope that this story has conveyed the rich texture of the development of calculus. It has deep roots in the ancient Middle East and Egypt. It was nurtured through its embrace by the Hellenistic Eastern Mediterranean, by India, by the Islamic Middle East, and by Europe. In the twentieth century, work on what was now known as analysis spread throughout the world. It is still growing.

In conducting the research for this book, I encountered some surprises. I had not been aware of the critical role played by Italy, assuming the creation of calculus was something that happened in northwestern Europe. Yet the debt to Italy makes sense. Most of the Islamic knowledge of mathematics was transmitted through Italy. In addition, here is where works of Euclid, Archimedes, and Pappus were translated into Latin, the *lingua franca* of the Renaissance. Galileo was enormously influential. As he demolished the assumptions of Aristotelian science, he recognized that the new foundation must lie within mathematics. The colleagues whom he mentored, Cavalieri and Torricelli, would do more than anyone else to set in motion the mathematical developments leading to calculus. Throughout the seventeenth century, Italy continued to be the primary country to which one traveled to learn about this subject. Napier went there to study mathematics, as did Barrow and Gregory. Wallis and a host of others learned the fundamentals of the subject by reading Torricelli.

I hope that I have done something to dispel the notion that calculus is purely the product of dead, white, male Europeans. It is true that most of what we teach as calculus arose from this small group. In fact, if you look through this book, you will see that every portrait of a contributor to the development of calculus is one of these dead, white, male Europeans. Part of this is due to the fact that before 1600 contributing to mathematics did not warrant immortalization in a contemporary bust or portrait.

Yet this limitation is mostly due to the fact that calculus itself could only emerge after two critical inventions: the transformation of algebra into a set of formal rules for operating on a refined symbolic notation and the marriage of algebra and geometry via Cartesian coordinates. As we have seen, algebra has ancient roots and was practiced at a high level of achievement in China, India, and the Middle East long before Europeans learned of it. But its true power was not unleashed until the decades just before and after 1600 when an efficient notation was developed and general rules for manipulating these symbols were formalized. Suddenly, broad categories of mathematical problems could be solved by translation into these symbols and application of these rules, eliminating hundreds of ad hoc procedures.

Analytic geometry was initially created to translate geometric problems into this powerful language of algebra where they could easily be solved. But the connection turned out to be even more useful in the other direction, translating algebraic relations into geometric curves, giving new insights into the interdependence of these variables via the areas and tangents that accompany these curves.

These inventions occurred only in Western Europe, particularly fertile ground because of the wealth of the societies that enabled support of philosophers, the infrastructure of a system of universities that provided centers for the exchange of ideas, and the intellectual freedom that encouraged questioning old certainties and the creation of new knowledge.[13] While understanding of these European creations did spread around the world, the transmission was slow. Those in other cultures and civilizations who built on what they learned were playing catch-up, rediscovering what was already known in Europe.

I included an account of Srinivasa Ramanujan in chapter 5 because he was arguably the first non-European to pick up existing work in analysis and carry it far beyond what any Europeans had accomplished. His story is particularly instructive because much of his labor required rediscovering what was already known and working through mistaken beliefs that had entrapped Europeans decades earlier. He would not remain alone for long. Today, mathematical breakthroughs are being made by people of all races and ethnicities.

For women, their scarcity in this story arises from the fact that until the late nineteenth century they were denied access to most European universities. Sofia Kovalevskaya, described in section 5.3 as the first woman appointed to a professorship at a European university, studied with Weierstrass at the University of Berlin. But, as a woman, she was never allowed to matriculate or even attend its classes. Weierstrass tutored her privately. After she wrote three papers, each of which Weierstrass considered worthy of recognition as Doctor of Philosophy, he convinced the University of Göttingen to confer on her this degree. While the situation has improved, today's women still face obstacles. They account for almost a third of the doctorates in the mathematical sciences awarded in the United States but represent only 16% of the mathematics faculty in U.S. research universities. It was not until 2014 that Maryam Mirzakhani (1977–2017) became the first woman to receive a Fields Medal, one of the two highest honors awarded in mathematics.[14]

Today we live in a rich age of mathematical discovery. Not all students have equal access to its beauties. That is a profound shame for there is nothing inherently European or male in this pursuit of the deep patterns of nature.

Notes

Preface

1. In modern notation, $\frac{1}{2} \times (\pi \times 2r) \times r$ or πr^2.
2. Grabiner, 1981.

Chapter 1. Accumulation

1. Dijksterhuis, 1956, p. 32.
2. The symbol π for this ratio was first used in the seventeenth or early eighteenth century—probably because it is the first letter of the Greek word for "perimeter"—and was popularized by Leonhard Euler.
3. Katz, 2009, p. 190.
4. By Islamic, I mean from the Islamic countries and cultures. Some of these philosophers were Jewish.
5. This uses the result, whose origins are lost in the mists of time, that $1 + 2 + \cdots + (n-1) = n(n-1)/2$.
6. In terms of p_k, $p_{k+1}(n) = \frac{n^{k+1}}{k+2} + \frac{k+1}{k+2}\left(np_k(n) - p_k(n-1) - p_k(n-2) - \cdots - p_k(1)\right)$. If $p_k(x) = a_k x^k + a_{k-1}x^{k-1} + \cdots + a_0$, then $p_k(n) + p_k(n-1) + \cdots + p_k(1) = a_k(n^k + (n-1)^k + \cdots + 1^k) + a_{k-1}(n^{k-1} + (n-1)^{k-1} + \cdots + 1^{k-1}) + \cdots + a_0(1 + 1 + \cdots + 1) = a_k(n^{k+1}/(k+1) + p_k(n)) + a_{k-1}(n^k/k + p_{k-1}(n)) + \cdots + a_0 \cdot n$, a polynomial in n of degree at most $k+1$.
7. Blaise Pascal's *Treatise on the Arithmetical Triangle* was published in 1665, popularizing these coefficients and forever attaching his name to them. He never claimed to have discovered them.
8. If we know that a polynomial in x has opposite signs at two consecutive integers, a and $a+1$, then there is a root between these integers. By substituting the binomial $a + y/10$ for x and using the binomial theorem to expand each power of this binomial, we obtain a polynomial in y that has a root between 0 and 10. Finding the integer b so that this polynomial has opposite signs at b and $b+1$, we know that the tenths digit of the root is b. Now substituting $b + z/10$ for y and expanding

again, we obtain a polynomial in z that can be used to find the hundredths digit of the root. This process can be continued indefinitely.

9. Heath, 1921, vol. 1, pp. 360–369.

10. Dijksterhuis, 1956, pp. 36–42.

11. Baron, 1969, pp. 91–96.

12. Baron, 1969, pp. 96–107.

13. Guldin did not mention the priority of Pappus. There has been considerable debate over whether Guldin plagiarized this result. Guldin must have read this result in Pappus's *Collection* several decades before he published it as his own—the work was too important and Guldin was too good a mathematician not to have read it. But Guldin also was very careful to properly attribute ideas he had gleaned from others. It is most likely that over the ensuing years he had forgotten that this particular result lay in the *Collection*.

14. Baron, 1969, p. 122.

15. Cavalieri used a parallelogram rather than a rectangle. I have chosen a rectangle to keep the picture a little simpler.

16. Although I have chosen to write this sum as $\sum_{0 \leq \ell \leq A}$, I am using this notation in an unconventional manner, indicating a summation over infinitely many lines that range in length from 0 to A.

17. Kirsti Andersen reports that Maximilien Marie, writing in the 1880s, suggested that if there were a prize for the world's most unreadable book, it should go to Cavalieri's 700-page *Geometria indivisibilibus* (Andersen, 1985, p. 294).

18. This reconstruction is based on Michael S. Mahoney's account (Mahoney, 1994).

19. Fermat never explicitly used induction, but it was implicit in many of his arguments and certainly applies here.

20. We do not know how Fermat proceeded. With the observation that $a_1 = k(k-1)/2$ and a little bit of work, it is possible to use equation (1.10) to prove Roberval's inequalities.

21. From Torricelli's *De solido hyperbolico acuto* as quoted in (Mancuso and Vailati, 1991, p. 53).

22. Cavalieri to Torricelli, Dec. 17, 1641, as quoted in (Mancuso and Vailati, 1991, p. 51).

23. Mancuso and Vailati, 1991, p. 58.

24. Heath, 1921, vol. 1, p. 276.

25. Clagett, 1959, p. 167.

26. Clagett, 1959, pp. 206–208.

27. Clagett, 1959, p. 236.

28. Clagett, 1959, pp. 277, 286–287.

29. Clagett, 1959, p. 332.

30. Clagett, 1959, p. 332.

31. Dijksterhuis, 1986, p. 331.
32. Clagett, 1959, p. 418
33. Heilbron, 2010.
34. Drake, 1978, p. 1.
35. "Calculus" is used here in its original meaning: a tool for computation.
36. Newton, 1687, p. 943. The original sentence, "Hypotheses non fingo," was translated by Motte as "I frame no hypotheses." I. Bernard Cohen has explained why this translation no longer adequately conveys Newton's intention (Newton, 1687, p. 274).
37. Galileo died on January 8, 1642. Newton was born on December 25, of the same year, which would appear to imply that Newton was born in the year of Galileo's death. But the dates are deceiving. From 1582 until 1752, Britain and Italy used different calendars. Britain was still on the old Julian calendar. Italy had adopted the new Gregorian calendar. January 8, 1642, in Italy corresponded to December 29, 1641, in Britain. December 25, 1642, in Britain corresponded to January 4, 1643, in Italy.
38. Van Schooten, 1649, pp. 203–205.
39. S. Chandrasekhar has translated large swathes of the *Principia* into the language of calculus (Chandrasekhar, 1995).
40. There is disagreement among scholars whether Newton proved that the inverse square law implies elliptical orbits. He did prove that an elliptical orbit implies the inverse square law of acceleration due to gravity. If the solution is unique, then the implication also goes the other way, but the uniqueness of the solution is not mentioned in the first edition, and only mentioned without elaboration in subsequent editions.

Chapter 2. Ratios of Change

1. These formulas were known to Ptolemy in second century CE Alexandria and probably traveled from there to Indian astronomers.
2. Plofker has given a detailed account of the derivation of equation (2.2) (Plofker, 2007, pp. 481–493).
3. Why 10^7? At this time, decimal fractions existed but were not in general use. To represent seven-digit accuracy on values of the sine function, one considered the lengths of the half-chords in circles of radius 10^7, which could then be represented as integers. Napier was working with logarithms of these half-chords.
4. See (Havil, 2014, pp. 100f) to see how Napier actually approached this.
5. The method for constructing these tables was ingenious. It is described in (Toeplitz, 2007, pp. 86f) and in detail in (Havil, 2014, pp. 107f).

6. Note that in the example in section 2.2, the fact that the base was chosen as 10 simplified dealing with integer powers of the base, because multiplication by powers of 10 is simply a matter of moving the decimal point.
7. The family name was Kauffman, but he changed it to Mercator.
8. Al-Khwarizmi, 1915, p. 34.
9. Al-Khwarizmi, 1915, p. 38.
10. Cardano, 1968, p. 11.
11. Smith, 1925, vol. 2, p. 430.
12. The problem does not require that these be perpendicular distances. They could be oblique. But we lose no generality in making the assumption that they are perpendicular.
13. Heath, 1921, vol. 2, p. 402.
14. Descartes, 1925.
15. Descartes, 1925, p. 95.
16. If the right-hand side is negative, there are no points that satisfy Apollonius's condition.
17. In modern notation, representing the points by their coordinates and working with directed distance, we can express this condition as $\sum_{i=1}^{n}(x - a_i) = 0$.
18. Boyer, 1959, p. 76.
19. Mahoney, 1994, p. 34.
20. Mahoney, 1994, pp. 147–150.
21. Descartes engaged in a vitriolic attack on Fermat, motivated in part by their simultaneous discovery of coordinate geometry. See (Mahoney, 1994, pp. 170–193) for a fascinating account of how and why this animosity unfolded.
22. Scriba, 1970.
23. Scriba, 1970, p. 26.
24. Scriba, 1970, pp. 29–30.
25. Scriba, 1970, p. 40.
26. Wallis, 2004, p. 15.
27. Wallis, 2004, p. xxiii.
28. Wallis, 2004, p. 22.
29. Wallis, 2004, p. 42.
30. See (Bressoud, 2007, pp. 271–277) or (Roy, 2011, pp. 28–33) for a description and discussion of his derivation of this result.
31. As I will explain in section 2.7, I prefer the full designation Fundamental Theorem of Integral Calculus over what has become more common today, the Fundamental Theorem of Calculus.
32. Fermat, isolated from the mainstream of mathematical work, would rediscover this formula two years later (Stedall, 2008, p. 102).

33. D. T. Whiteside identified this as one of the two criteria necessary to any working definition of calculus, "First, that differentiation and integration be seen as inverse procedures; and, secondly that both be defined with respect to an adequate algorithmic technique" (Whiteside, 1962, p. 365).

34. While the relationship that lies at the heart of this theorem was known to Newton and Leibniz and is implicit in the work of Oresme in the fourteenth century that recognized the area under the graph of velocity as representing distance, the modern formulation of this result is due to Cauchy in the 1820s. In Cauchy, it appears as part of his justification for the definition of the definite integral as a limit of sums. He never identifies it as a theorem. The earliest record I can find of anyone referring to this as a theorem was Paul du Bois-Reymond in 1876, who felt the need to prove it in an appendix to a paper on Fourier series. He called it the Fundamental Theorem of Integral Calculus, a name that stuck until the popular calculus textbooks of the 1960s shortened it to the Fundamental Theorem of Calculus.

35. Apparently, Newton gave it to his college roommate, Wickins, to be passed on to Collins. We do not know whether Collins ever received it. Some time in the eighteenth century it wound up back in the Cambridge library.

36. It was preceded by the *Philosophical Transactions* of the Royal Society in England and the *Journal des Sçavans* in France, both founded in 1665, and the *Giornale de'letterati*, started in 1668 in Italy.

37. An English translation of this article, "A new method for maxima and minima as well as tangents which is neither impeded by fractional nor irrational quantities, and a remarkable type of calculus for them," is contained in (Struik, 1996, pp. 271–280).

38. The modern French spelling of the word for hospital is hôpital. This is a modern contraction of the older spelling that coincides with the English spelling, and it is this older spelling that the marquis used when he wrote his name. When pronouncing his name, the "s" is silent.

39. Euler, 2008.

40. For a function of several variables such as $u(x, y, a, t)$, the partial derivative of u with respect to x, written $\partial u / \partial x$, describes the rate at which u changes when only x changes.

41. Doubling the frequency raises the pitch by an octave. A 50% increase raises it by a major fifth, 25% by a major third. To multiply by 5, we double twice, then increase that amount by 25%.

42. It did not hurt that Parisians viewed North America as a vast wilderness from which Franklin had emerged as a "child of nature," untainted by modern society. He played this up by wearing a coonskin cap.

43. If you really want to know: Given a current flow described by the vector $\langle j_1, j_2, j_3 \rangle$ and the corresponding magnetic field given by $\langle B_1, B_2, B_3 \rangle$, the result discovered

by Biot and Savart, now known as Ampère's law, states that

$$\left\langle \frac{\partial B_3}{\partial y} - \frac{\partial B_2}{\partial z}, \frac{\partial B_1}{\partial z} - \frac{\partial B_3}{\partial x}, \frac{\partial B_2}{\partial x} - \frac{\partial B_1}{\partial y} \right\rangle = \mu \langle j_1, j_2, j_3 \rangle,$$

where μ is a constant known as the *magnetic permeability*.

44. In this case, the governing partial differential equation is

$$\left\langle \frac{\partial E_3}{\partial y} - \frac{\partial E_2}{\partial z}, \frac{\partial E_1}{\partial z} - \frac{\partial E_3}{\partial x}, \frac{\partial E_2}{\partial x} - \frac{\partial E_1}{\partial y} \right\rangle + \left\langle \frac{\partial B_1}{\partial t}, \frac{\partial B_1}{\partial t}, \frac{\partial B_1}{\partial t} \right\rangle = 0,$$

where $\langle E_1, E_2, E_3 \rangle$ is the electrostatic force that can push electrons around the circuit.

Chapter 3. Sequences of Partial Sums

1. Back in the fourteenth century, Nicole Oresme had also demonstrated that this series grows without bound.
2. Ferraro, 2008, pp. 41–42.
3. He actually used $B, C, D, \ldots$ for the coefficients.
4. Called the *interpolating polynomial* because it can be used to interpolate intermediate values.
5. The general theorem for representing a function in terms of its zeros is due to Karl Weierstrass (1895).
6. Because $\lim_{x \to 0} \frac{\sin x}{x} = 1$, defining $\frac{\sin x}{x}$ to be 1 at $x = 0$ produces a function that is continuous at 0.
7. If we equate the coefficients of x^4, we see that $\sum_{i<j} i^{-2} j^{-2} = \pi^4 / 120$. This enabled Euler to sum the reciprocals of fourth powers,

$$\sum_{i=1}^{\infty} \frac{1}{i^4} = \left(\sum_{i=1}^{\infty} \frac{1}{i^2} \right)^2 - 2 \sum_{i<j} \frac{1}{i^2 j^2} = \frac{\pi^4}{36} - \frac{\pi^4}{60} = \frac{\pi^4}{90}.$$

This can be extended to the reciprocals of any positive even power.

8. D'Alembert, 1768, pp. 173–174.
9. Reprinted by J. A. Serret in 1847 (Lagrange, 1847).
10. Bers, 1967.
11. As translated in (Smith, 1982, p. 7).
12. This became the model for the U.S. Military Academy at West Point, which accepted the assumption that the best academic preparation for a future officer is an engineering education.

13. One of the first women to make significant contributions to mathematics in western Europe, Germain (1776–1831) worked in number theory, where she is best known for her contributions toward the solution of Fermat's Last Theorem as well as her prize-winning essay on the mathematics of elastic surfaces.

Chapter 4. The Algebra of Inequalities

1. Today known as Oslo, Norway.
2. See (Gauss, 1812). A hypergeometric series is any series, $\sum a_n$, for which a_{n+1}/a_n is a rational function of n, that is to say, a ratio of two polynomials in n. For example, the series for sine, $\sum_{n=1}^{\infty} x^{2n-1}/(2n-1)!$, is hypergeometric because

$$\frac{x^{2n+1}}{(2n+1)!} \frac{(2n-1)!}{x^{2n-1}} = \frac{x^2}{(2n)(2n+1)}$$

is a constant with respect to n, namely x^2, divided by a polynomial in n. Outside of a calculus class, it is unlikely you will ever encounter a Taylor series that is not hypergeometric. See (Bressoud, 2007, pp. 149–153) for details of Gauss's theorem.
3. Wallis, 2004, p. 27.
4. Grabiner, 1981, p. 84.
5. An event that, fortunately, never happened, Napoleon having turned his attention to the invasion of Russia.
6. Strictly speaking, he was tied for first. Upon learning that his friend Jacques Binet (1786–1856) had discovered the same result, they arranged to submit both of their manuscripts to the Institut de France on the same day.
7. Cauchy, 1821.
8. Cauchy, 1823.
9. An ϵ–δ proof only requires showing that there is some $\delta > 0$ that will work for any given $\epsilon > 0$. Because

$$1 - \frac{x^2}{2} < \cos x < 1$$

when $x \neq 0$, we can meet the required tolerance by setting $\delta = \sqrt{2\epsilon}$. Note that since $(\sin x)/x$ is not defined when $x = 0$, we need the lower inequality, $0 < |x - 0|$.
10. To be precise, by the greatest value we mean the smallest value that is greater than or equal to every value of f on the set, technically known as the *supremum*, and the least value means the largest number that is less than or equal to every value of f, technically known as the *infimum*.
11. Note that for each fraction a/b, we have $a^2 - 2b^2 = \pm 1$, so that $(a/b)^2 - 2 = \pm 1/b^2$. If (a, b) satisfies this equation, then so does $(a + 2b, a + b)$. As the denominator, b, increases, these fractions are getting closer to $\sqrt{2}$.

12. We have shown that if every Cauchy sequence converges, then every bounded increasing sequence converges. In the other direction, it is only necessary to show that every Cauchy sequence has either an infinite increasing subsequence or an infinite decreasing subsequence, and then to establish that the limit of this subsequence must also be the limit of the Cauchy sequence.

13. Cauchy, 1821, p. 34

14. A proof of this result can be found in (Bressoud, 2007, p. 229).

15. Abel, 1826, pp. 224–225.

16. Lützen, 2003.

Chapter 5. Analysis

1. Meaning that if $\pi(n)$ is the number of primes less than or equal to n, then

$$\lim_{n \to \infty} \frac{\pi(n)}{n/\ln n} = 1.$$

2. An explanation of why it has these properties can be found in (Bressoud, 2007, pp. 252–254).

3. See (Bressoud, 2007, p. 112) for a proof.

4. Volterra, 1881.

5. For a discussion of the Henstock integral, see (Bressoud, 2008, pp. 291–296).

6. Bell, 1937, p. 406.

7. According to the Mathematics Genealogy Project (www.genealogy.math.ndsu. nodak.edu), he had 42 doctoral students. Almost 32,000 mathematicians can trace their mathematical lineage back to him.

8. Dunham, 2005, pp. 128–148.

9. Actually, a teacher in a *gymnasium*, an academic secondary school designed to prepare students for university.

10. This was conjectured by Fermat and proven in 1994 by Andrew Wiles: For $n \geq 3$, the equation $a^n + b^n = c^n$ does not have any solutions in positive integers.

11. Degree 4.

12. Explanations of the role of the elliptic integral in the study of pendular motion can be found in (Toeplitz, 2007, pp. 138–142) and (McKean and Moll, 1999, pp. 58–59).

13. Gauss, 1929.

14. The hyperbolic cosine and sine are, respectively, the even and odd parts of the exponential function, $e^x = \cosh x + \sinh x$, where $\cosh x = (e^x + e^{-x})/2$ and $\sinh x = (e^x - e^{-x})/2$. They satisfy $\cosh^2 x - \sinh^2 x = 1$, $\frac{d}{dx}\sinh x = \cosh x$, and

$\frac{d}{dx}\cosh x = \sinh x$. With the substitution $x = \sinh u$, we have

$$\int_0^t \frac{dx}{\sqrt{1+x^2}}\,dx = \int_0^{\text{arcsinh}\,t} \frac{\cosh u\,du}{\sqrt{1+\sinh^2 u}} = \int_0^{\text{arcsinh}\,t} \frac{\cosh u\,du}{\cosh u} = \text{arcsinh } t.$$

15. Today called Chennai.
16. Finitely many points of D would leave open intervals which must contain more points of D.
17. It appears in Cantor's collected works, (Cantor, 1932, pp. 115–118).
18. For example, see (Courant and Robbins, 1978).
19. Dunham, 2005, pp. 161–164.
20. In section 5.5, we will explore the question of the existence of a set that does not have a measure.
21. We do need to worry about decimals with infinitely repeating 9's, but there are only countably many such exceptions.
22. Just as 2^k is the number of mappings from a k-element set to a two-element set.
23. One line of attack is to expand the Zermelo-Fraenkel axioms by adding additional assumptions that seem necessary and would settle the continuum hypothesis one way or the other. None of these attempts have been entirely satisfactory.
24. For details of its construction, see (Bressoud, 2008, pp. 150–151).
25. For those familiar with more advanced mathematics, these included: every proper ideal of a commutative ring can be extended to a maximal ideal, every vector space has a basis, and every Hilbert space has an orthonormal basis.
26. Although the number of pieces is finite, it would need to be very large because the total reconstituted volume cannot be larger than the volume of the pea multiplied by the number of pieces.

Appendix. Reflections on the Teaching of Calculus

1. Martzloff, 1997, pp. 14–15.
2. See "Project DIRACC: Developing and Investigating a Rigorous Approach to Conceptual Calculus." http://patthompson.net/ThompsonCalc.
3. See "Project CLEAR Calculus: Coherent Labs to Enhance Accessible and Rigorous Calculus." http://clearcalculus.okstate.edu/about-clear-calculus.
4. Note that the graph looks suspiciously like a parabola. If you substitute the power series expansion for $e^{-t/25}$, you will see why.
5. Macalester's only calculus sequence emphasizes calculus as a tool for modeling dynamical systems throughout all three semesters. It works with functions of several variables from the beginning of the first semester.

6. In the epilogue to (Oehrtman et al., 2014), the authors discuss student difficulties moving from sequences and series with numerical values to sequences and series of functions, often confusing $\lim_{n\to\infty}$ with $\lim_{x\to\infty}$.
7. Swinyard, 2011.
8. Swinyard and Larsen, 2012.
9. Tall and Vinner, 1981.
10. Oehrtman, 2009.
11. Oehrtman et al. 2008.
12. Lax and Terrell, 2014.

The Last Word

13. It is not surprising that tiny Netherlands—combining wealth, a strong academic tradition, and complete freedom from the Church—would play such an important role in the history of calculus.
14. The Fields Medal, first awarded in 1936 and restricted to work done before the age of forty, had been given to 56 people through 2014. Equally prestigious but given for lifetime achievement in mathematics, the Abel Prize was first awarded in 2003. As of this writing, it has never been given to a woman.

Bibliography

Abel, N. H. (1826). Recherches sur la série $1 + \frac{m}{1}x + \frac{m(m-1)}{1.2}x^2 + \frac{m(m-1)(m-2)}{1.2.3}x^3 + \cdots$. *Journal für die reine und angewandte mathematik* 1, 219–250.

al-Khwarizmi, M. (1915). *Robert of Chester's Latin Translation of the Algebra of Al-Khowarizmi*. New York, NY: Macmillan Company. English translation by Louis Charles Karpinski.

Andersen, K. (1985). Cavalieri's method of indivisibles. *Archive for History of Exact Sciences* 31, 291–367.

Barbeau, E. J., and P. J. Leah (1976). Euler's 1760 paper on divergent series. *Historia Mathematica* 3, 141–160.

Baron, M. E. (1969). *The Origins of the Infinitesimal Calculus*. Oxford: Pergamon Press.

Bell, E. T. (1937). *Men of Mathematics*. New York, NY: Simon and Schuster.

Bers, L. (1967). On avoiding the mean value theorem. *American Mathematical Monthly* 74, 583.

Boyer, C. B. (1959). *The History of the Calculus and its Conceptual Development*. New York: Dover. Reprint of *The Concepts of the Calculus*, New York, NY: Hafner, 1949.

Bressoud, D. M. (2007). *A Radical Approach to Real Analysis* (2nd ed.). Washington, DC: Mathematical Association of America.

Bressoud, D. M. (2008). *A Radical Approach to Lebesgue's Theory of Integration*. Cambridge: Cambridge University Press.

Cantor, G. (1932). Über eine eigenschaft des inbegriffes aller reellen algebraischen zahlen. In A. Fraenkel (Ed.), *Georg Cantor Gesammelte Abhandlungen*, pp. 115–118. Berlin: Verlag von Julius Springer.

Cardano, G. (1968). *The Great Art or The Rules of Algebra*. Cambridge, MA: MIT Press. Translated and edited by T. Richard Witmer.

Cauchy, A.-L. (1821). *Cours d'Analyse de L'École Royale Polytechnique*. Paris: L'Imprimerie Royale.

Cauchy, A.-L. (1823). *Résumé des leçons donnés a L'École Royale Polytechnique sur le Calcul Infinitésimal*, vol. 1. Paris: L'Imprimerie Royale.

Chandrasekhar, S. (1995). *Newton's Principia for the Common Reader*. Oxford: Oxford University Press.

Child, J. M. (1920). *The Early Mathematical Manuscripts of Leibniz*. Chicago, IL: Open Court Publishing.

Clagett, M. (1959). *The Science of Mechanics in the Middle Ages*. Madison, WI: University of Wisconsin Press.

Copernicus, N. (1543). *De revolutionibus orbium cœlestium*. Nuremberg: Ioh. Petreium. Translated by Edward Rosen, 1978; 1999 CD published by Octavo.

Courant, R., and H. E. Robbins (1978). *What is Mathematics?* Oxford, England: Oxford University Press.

d'Alembert, J. (1768). Réflexions sur les suites et sur les racines imaginaires. *Opuscules mathématiques* 5, 171–215.

Descartes, R. (1925). *The Geometry of René Descartes*. Chicago: Open Court Publishing. Translated by David Eugene Smith and Marcia L. Latham, with a facsimile of the first edition, 1637.

Dijksterhuis, E. J. (1956). *Archimedes*. Princeton, NJ: Princeton University Press. Translated by C. Dikshoorn.

Dijksterhuis, E. J. (1986). *The Mechanization of the World Picture: Pythagoras to Newton*. Princeton, NJ: Princeton University Press. Translated by C. Dikshoorn.

Drake, S. (1978). *Galileo at Work: His Scientific Biography*. Chicago, IL: University of Chicago Press.

Dunham, W. (2005). *The Calculus Gallery: Masterpieces from Newton to Lebesgue*. Princeton, NJ: Princeton University Press.

Euclid (1956). *The Thirteen Books of Euclid's Elements* (2nd ed.). New York: Dover. Translated by Thomas L. Heath.

Euler, L. (1988). *Introduction to Analysis of the Infinite*, Volume 1. New York: Springer Verlag. Translated by John D. Blanton.

Euler, L. (2000). *Foundations of Differential Calculus*. New York: Springer Verlag. Translated by John D. Blanton.

Euler, L. (2008). Principles of the motion of fluids. *Physica D: Nonlinear Phenomena* 237. English adaptation by Walter Pauls of Euler's memoir "Principia motus fluidorum" (Euler, 1756–1757).

Ferraro, G. (2008). *The Rise and Development of the Theory of Series Up to the Early 1820s*. New York, NY: Springer Verlag. Sources and Studies in the History of Mathematics and Physical Sciences.

Feynman, R., M. Sands, and R. B. Leighton (1964). *The Feynman Lectures on Physics*, Vol. 2. Reading, MA: Addison-Wesley.

Galilei, G. (1638). *Dialogues Concerning Two New Sciences* (trans. H. Crew and A. de Salvio) (2nd ed.). New York: Dover. 1954; reprint of New York: Macmillan, 1914.

Gauss, C. F. (1812). *Disquisitiones generales circa seriem infinitam . . . .* Göttingen: Societas Regia Scientiarum Gottingensis.

Gauss, C. F. (1870–1929). Elegantiores integralis $\int (1 - x^4)^{-1/2} dx$ propruetates et de curva lemniscata. In *Werke*, pp. 404–432. Göttingen: Königliche Gesellschaft der Wissenschaft.

Grabiner, J. V. (1981). *The Origins of Cauchy's Rigorous Calculus*. Cambridge, MA: MIT Press.

Havil, J. (2014). *John Napier: Life, Logarithms, and Legacy*. Princeton, NJ: Princeton University Press.

Heath, T. (1921). *A History of Greek Mathematics*. Oxford: Clarendon Press. Reprinted by Dover, 1981.

Heilbron, J. (2010). *Galileo*. Oxford, England: Oxford University Press. Originally published 1921.

Katz, V. J. (2009). *A History of Mathematics: An Introduction* (3rd ed.). Boston, MA: Addison-Wesley.

Lagrange, J.-L. (1847). *Théorie des fonctions analytiques* (3rd ed.). Paris: Bachelier.

Lax, P. D., and M. S. Terrell (2014). *Calculus with Applications* (2nd ed.). New York: Springer-Verlag.

Lützen, J. (2003). The foundation of analysis in the 19th century. In H. N. Jahnke (Ed.), *A History of Analysis*, pp. 155–195. Providence, RI: American Mathematical Society.

Mahoney, M. S. (1994). *The Mathematical Career of Pierre de Fermat* (2nd ed.). Princeton, NJ: Princeton University Press.

Mancuso, P., and E. Vailati (1991). Torricelli's infinitely long solid and its philosophical reception in the seventeenth century. *Isis* 82, 50–70.

Martzloff, J.-C. (1997). *A History of Chinese Mathematics*. Berlin: Springer Verlag. Translated by Stephen S. Wilson from the French original.

McKean, H., and V. Moll (1999). *Elliptic Curves: Function Theory, Geometry, Arithmetic*. Cambridge, England: Cambridge University Press.

Newton, I. (1666). The October 1666 tract on fluxions. In D. T. Whiteside (Ed.), *The Mathematical Papers of Isaac Newton*, vol. 1, 1664–1666, pp. 400–448. Cambridge: Cambridge University Press.

Newton, I. (1687). *The Principia: Mathematical Principles of Natural Philosophy*. Berkeley, CA: University of California Press. Translation by I. Bernard Cohen and Anne Whitman, originally published 1687.

NOVA (2003). *Infinite Secrets: The Genius of Archimedes*. DVD, WGBH Boston.

Oehrtman, M. (2009). Collapsing dimensions, physical limitation, and other student metaphors for limit concepts. *Journal for Research in Mathematics Education* 40, 396–426.

Oehrtman, M., M. Carlson, and P. W. Thomson (2008). Foundational reasoning abilities that promote coherence in students' function understanding. In M. P. Carlson and C. Rasmussen (Eds.), *Making the Connection: Research and Teaching in*

Undergraduate Mathematics Education, pp. 27–41. Washington, DC: Mathematical Association of America.

Oehrtman, M., C. Swinyard, and J. Martin (2014). Problems and solutions in students' reinvention of a definition for sequence convergence. *Journal of Mathematical Behavior* 33, 131–148.

Ore, O. (1974). *Niels Henrik Abel, Mathematician Extraordinary*. Minneapolis, MN: University of Minnesota Press.

Plofker, K. (2007). Mathematics in India. In V. J. Katz (Ed.), *The Mathematics of Egypt, Mesopotamia, China, India, and Islam: A Sourcebook*, pp. 385–514. Princeton, NJ: Princeton University Press.

Poincaré, H. (1889). La logique and l'intuition dans la science mathématique et dans énseignement. *L'Enseignement mathématique* 11, 157–162.

Roy, R. (2011). *Sources in the Development of Mathematics: Infinite Series and Products from the Fifteenth to the Twenty-first Century*. Cambridge: Cambridge University Press.

Scriba, C. J. (1970). The autobiography of John Wallis, F.R.S. *Notes and Records of the Royal Society of London* 25, 17–46.

Smith, D. E. (1923,1925). *History of Mathematics*. Boston, MA: Ginn and Company.

Smith, R. J. (1982). *The École Normale Supérieure and the Third Republic*. Albany, NY: State University of New York Press.

Stedall, J. (2008). *Mathematics Emerging: A Sourcebook 1540–1900*. Oxford: Oxford University Press.

Struik, D. J. (1996). *A Source Book in Mathematics: 1200–1800*. Berlin: Springer-Verlag.

Stubhaug, A. (1996). *Niels Henrik Abel and His Times: Called Too Soon by Flames Afar*. Berlin: Springer Verlag. Translated by Richard H. Daly.

Swinyard, C. (2011). Reinventing the formal definition of limit: The case of Amy and Mike. *Journal of Mathematical Behavior* 30, 93–114.

Swinyard, C., and S. Larsen (2012). Coming to understand the formal definition of limit: Insights gained from engaging students in reinvention. *Journal for Research in Mathematics Education* 43, 465–493.

Tall, D., and S. Vinner (1981). Concept image and concept definition in mathematics with particular reference to limits and continuity. *Educational Studies in Mathematics* 12, 151–169.

Toeplitz, O. (2007). *The Calculus: A Genetic Approach*. Chicago, IL: University of Chicago Press.

Van Schooten, F. (1649). *Geometria à Renato Des Cartes*. Leiden: Ioannis Maire.

Volterra, V. (1881). Alcune osservazioni sulle funzioni puteggiate discontinue. *Giornale di Matematiche* 19, 76–86.

Wallis, J. (2004). *The Arithmetic of Infinitesimals*. New York, NY: Springer-Verlag. Translated with an introduction by Jacqueline A. Stedall.

Wapner, L. M. (2005). *The Pea and the Sun: A Mathematical Paradox*. Wellesley, MA: A K Peters.

Weierstrass, K. (1895). Zur Theorie der eindeutigen analytischen Funktionen. In *Mathematische werke von Karl Weierstrass*, vol. 2, pp. 77–124. Berlin: Mayer and Müller.

Whiteside, D. T. (1962). Patterns of mathematical thought in the later seventeenth century. *Archive for History of Exact Sciences* 1, 179–388.

INDEX

Image Credits

Figure 1.17. Portrait of Pierre de Fermat from "Oevres de Fermat, vol. 1 (Gauthier-Villars, Paris, MDCCCXCI)." University of Rochester, courtesy AIP Emilio Segrè Visual Archives.

Figure 1.22. Portrait of Galileo Galilei painted by Justus Sustermans, R. Galleria Uffizi, lithograph by Photographische Gesellschaft in Berlin. Courtesy AIP Emilio Segrè Visual Archives, W.F. Meggers Collection.

Figure 1.24. Sir Isaac Newton. From *The Calculus Gallery*, by William Dunham, 2008. Princeton University Press.

Figure 1.25. Christiaan Huygens. ©Huntington Library, courtesy AIP Emilio Segrè Visual Archives, Burndy Library Collection.

Figure 1.29. From Newton's *Principia*, 1687, as reproduced in *The Principia*, by Isaac Newton, and translated by I. Bernard Cohen and Anne Whitman, ©1999 by the Regents of the University of California. Published by the University of California Press.

Figure 1.30. The title page of Newton's *Philosophiae Naturalis Principia Mathematica* (first issue, first edition, London, 1687). New York Public Library/Science Source.

Figure 1.31. From Newton's *Principia*, 1687, as reproduced in *The Principia*, by Isaac Newton, and translated by I. Bernard Cohen and Anne Whitman, ©1999 by the Regents of the University of California. Published by the University of California Press.

Figure 2.4. John Napier. Copyright unknown. Widely available online. Source: http://www-history.mcs.st-and.ac.uk/history/PictDisplay/Napier.html. Accessed August 6, 2018.

Figure 2.8. François Viète. Copyright unknown. Widely available online. Source: https://commons.wikimedia.org/wiki/File:Francois_Viete.jpg. Accessed August 6, 2018.

Figure 2.9. René Descartes. Engraving of René Descartes by W. Holl from original by Francis Hals in the gallery of the Louvre. AIP Emilio Segrè Visual Archives.

Figure 2.15. Portrait of John Wallis. AIP Emilio Segrè Visual Archives, E. Scott Barr Collection.

Figure 2.16. James Gregory. Portrait in oils of James Gregory, mathematician and inventor of the reflecting telescope, attributed to Richard Waitt, 1708–1732. ©National Museums Scotland.

Figure 2.18. Gottfried Leibniz. From *The Calculus Gallery*, by William Dunham, 2008. Princeton University Press.

Figure 2.19. Leonhard Euler. From *The Calculus Gallery*, by William Dunham, 2008. Princeton University Press.

Figure 2.23. James Clerk Maxwell. Copyright unknown. Widely available online. Source: https://www.biography.com/people/james-c-maxwell-9403463. Accessed April 10, 2018.

Figure 3.2. Portrait of Jean le Rond d'Alembert. AIP Emilio Segrè Visual Archives, E. Scott Barr Collection.

Figure 3.3. Engraved portrait of Joseph Louis Lagrange. AIP Emilio Segrè Visual Archives, E. Scott Barr Collection.

Figure 3.4. Portrait of Joseph Fourier. Watercolor by Julien-Léopold Boilly (1820). Album de 73 Portraits-Charge Aquarelle's des Membres de l'Institute (watercolor portrait #29). Bibliotheque de l'Institut de France. ©RMN-Grand Palais / Art Resource, NY.

Figure 4.1. Augustin-Louis Cauchy. From *The Calculus Gallery*, by William Dunham, 2008. Princeton University Press.

Figure 5.6. Karl Theodor Wilhelm Weierstrass. From *The Calculus Gallery*, by William Dunham, 2008. Princeton University Press.

The following figures were generated using *Mathematica*: 2.22, 3.1, 3.5, 3.6, 5.1, 5.2, 5.3, 5.4, 5.5, 5.7, A.1, and A.2.